Experimental Methods

Experimental Methods

W. Bolton

Newnes a division of Butterworth-Heinemann
Linacre House, Jordan Hill, Oxford OX2 8DP
A division of Reed Educational & Professional Publishing Ltd

A member of the Reed Elsevier plc group

OXFORD BOSTON JOHANNESBURG
MELBOURNE NEW DELHI SINGAPORE

First published 1996

© W. Bolton 1996

All rights reserved. No part of this publication
may be reproduced in any material form (including
photocopying or storing in any medium by electronic
means and whether or not transiently or incidentally
to some other use of this publication) without the
written permission of the copyright holder except in
accordance with the provisions of the Copyright,
Designs and Patents Act 1988 or under the terms of a
licence issued by the Copyright Licensing Agency Ltd,
90 Tottenham Court Road, London, England W1P 9HE.
Applications for the copyright holder's written permission
to reproduce any part of this publication should be addressed
to the publishers

British Library Cataloguing in Publication Data
A catalogue record for this book is available from the British Library
ISBN 0 7506 2953 3

Library of Congress Cataloguing in Publication Data
A catalogue record for this book is available from the Library of Congress

Printed and bound in Great Britain by
Martins the Printers Ltd, Berwick upon Tweed

Contents

		Preface	vii
1 Experiments	1.1	Stages in experimental work	1
	1.2	Record keeping	3
	1.3	Report writing	4
2 Experimental data	2.1	Units	11
	2.2	Significant figures	19
		Problems	22
3 Graphs	3.1	Plotting graphs	24
	3.2	Straight line graphs	30
	3.3	Linearising equations	33
	3.4	Areas under graphs	40
		Problems	41
4 Experimental errors	4.1	Why estimate errors?	46
	4.2	Sources of error	47
	4.3	Random and systematic errors	51
	4.4	The mean value and its error	51
	4.5	Combining errors	56
		Problems	67
5 Statistics and data	5.1	Distributions	70
	5.2	Standard deviation	75
	5.3	Standard error of the mean	81
	5.4	Normal distribution	84
	5.5	t-distribution	88
	5.6	Difference between two means	91
		Problems	93
6 Least squares	6.1	The method of least squares	97
	6.2	Errors	103
	6.3	Weighting of results	107
		Problems	112
7 Spreadsheets	7.1	Spreadsheets	115
	7.2	Spreadsheets and charts	124
		Problems	126

Appendix A: Experiments	127
Appendix B: Further reading	132
Answers	133
Index	135

Preface

This book is aimed at those beginning an undergraduate course or higher technician course in engineering or the physical sciences and who have to complete a laboratory course as part of their studies.

The aim of this book is to enable the reader to:

- Understand the reasons for keeping a laboratory book/diary/journal and the types of information that should be entered in it.
- Write suitable laboratory reports.
- Handle units.
- Understand the relevance of significant figures and rounding.
- Plot graphs, linearising equations where necessary.
- Recognise sources of error in experimental work.
- Determine the accuracy of experimental results.
- Use the least squares method of fitting experimental data to the straight line equation.
- Use spreadsheets in experimental work.

In general, the chapters include worked examples and problems, answers to all problems being given at the end of the book. A small number of simple experiments, which can be used to practise the principles introduced in the book, are included in Appendix A.

W. Bolton

1 Experiments

Engineers and scientists tend to spend a lot of time carrying out experiments. Experiments enable theories to be tested, relationships to be determined, quantities to be measured, answers obtained to questions of the form – 'What happens if?' and, in courses of laboratory work, might also be used to develop familiarity with apparatus and provide training in how to do experiments. For example, there might be measurements of the current through a resistor and the voltage across it to enable the relationship between the two quantities to be determined, a measurement of the speed of light, a determination of the tensile properties of a sample of a material by the use of a tensile test machine, the development of a circuit using an operational amplifier as an astable multivibrator with a frequency which is to depend on the temperature, an investigation of the changes in silver ion concentration in solution when different reagents are used, etc.

This chapter is a consideration of the general stages involved in carrying out experiments, the maintaining of records and the writing of reports of experimental work.

1.1 Stages in experimental work

The stages that are likely to be involved in experimental work are:

1. *Aim*
   The aim of the experiment needs to be defined and perfectly clear before any further work starts. What is the purpose of the experiment? What is to be found out? For example, the aim of an electrical experiment might be to find out whether a particular resistor obeys Ohm's law. Another experiment might be to find a value for the thermal conductivity of some material. Another experiment might be to investigate the chemical reaction between iodine and aqueous alcohol and determine the factors affecting the rate of the reaction.

2. *Plan*
   When the aim is clear, the experiment needs to be planned. This means making decisions about such matters as what measurements are to be made, how the measurements are to be made and what instruments are needed. Thus with an experiment to find out whether a resistor obeys Ohm's law the measurements to be made are current and voltage. What size currents and voltage are likely to be involved and so what meters should be selected?

3. *Preparations*
   Once the experiment is planned, preparations must be made to carry out the experiment. This involves collecting and assembling the required instruments and making certain you know how to operate them. For

example, if the experiment requires the use of an instrument with a vernier, do you know how to read such a scale? In carrying out the experiment, are there any health and safety factors which need to be taken into account? For example, will you need a fume cupboard or perhaps may care need to be exercised because of high voltages?

4 *Preliminary experiment*
In some cases a preliminary experiment might be needed to find out whether the method you propose or the instruments you have selected are suitable. For example, with the use of meters such as ammeters and voltmeters, a preliminary experiment might be used to determine whether the current and voltage ranges of the instruments are adequate to enable the full range of measurements to be made.

5 *Doing the experiment and making measurements*
The next stage is to carry out the experiment, making such measurements as are required. Steps should be taken to minimise errors by eliminating systematic errors (a systematic error is one that is constant through a set of readings, see Chapter 4), e.g. adjusting an ammeter so that it reads zero when there is no current, and techniques adopted to reduce the impact of random errors, e.g. taking a series of measurements and using the mean value (see Chapter 4). All measurements and details of the instruments used should be recorded (see section 1.2). For example, in the measurement of a temperature the instrument used needs to be recorded because it could have an impact on the data obtained. A mercury-in-glass thermometer, because it is relatively slow reacting, is likely to lag behind a fast- changing temperature change and so affect the results obtained, whereas a thermocouple is much faster reacting. Thus the temperature readings indicated by these could differ.

6 *Repeating measurements*
Often there can be a need to repeat some measurements in order to verify that the first set of results obtained can be reproduced and can thus be relied on. It is often worthwhile doing at least some preliminary analysis of the data before putting all the equipment away in order to check whether there is some oddity about a particular measurement or set of measurements and repetition is necessary. For example, with a multi-range ammeter you might have misread the range for some readings.

7 *Analysis of the data*
Analyse the data to find out what the data tells you, taking into account the accuracy of measurements, and estimate the accuracy of the final result (see Chapter 4).

8 *Report*
Finally there needs to be a report of the experiment in which the findings are communicated to others (see section 1.3).

1.2 Record keeping A *laboratory notebook* should be kept in which a permanent record of experiments is kept. Scraps of paper are not suitable, they can be easily lost. The terms *laboratory diary* and *laboratory journal* are often used in order to indicate that the record is made directly during the carrying out of experiments. Every detail of an experiment is recorded, whether at the time it seems important or not. It may not be until later when a report is being written, or someone queries your results or interpretation, that the significance of a particular piece of information becomes apparent. The information in the notebook is the basis on which reports are written and results analysed.

The following is the type of information that may be entered in a laboratory notebook:

1 The date on which the experiment was carried out.

2 The title of the experiment.

3 The aims of the experiment.

4 Details of the apparatus. This might include a listing of the serial numbers of instruments, so that, if there are queries at a later date, it is possible to recheck the instruments used. It should certainly indicate the ranges of instruments, the quoted accuracy and other specification data, e.g. the resistance of an ammeter or voltmeter.

5 Sketches of circuits and apparatus. Simple labelled line diagrams are generally all that is required to adequately record details of circuits and apparatus. Combined with a few words of explanation, a diagram is often the most effective way of describing apparatus or the principles of an experiment.

6 Record details of the experimental method used. If the method is given as a set of instructions then you may be able to 'cut and paste' the information in your notebook, adding any extra details that occur during your carrying out of the experiment. You need notes about everything you do, including notes about any difficulties experienced or unusual observations or effects.

7 Record all observations and readings directly into your notebook at the time they occur. Where possible, readings should be recorded in tables with columns headed with the name and unit. The table format is compact and easy to follow. Estimates should be made, and recorded, of the experimental errors associated with readings. For example, in the case of a reading of a scale you could list the interval between graduations and give an estimate of the reading error (see Chapter 4).

8 Graphs are a good way of enabling relationships between quantities to be seen. A rough graph can usefully be plotted during an experiment and might disclose that the relationship is perhaps not the one expected

and so consideration needs to be given as to whether there is some unforeseen error occurring. It might also indicate that the readings being taken are going to be all grouped together in one region of a graph and thus readings should be taken at other values to 'fill the gaps'. It might also indicate that a particular measurement appears to be 'off-line' and so needs checking.

9   Record details of all equations used and calculations that are made. You might make an error in your calculations and thus the more detail you record the easier it will be to check. You might, for example, use a spreadsheet program and should thus indicate the formulas used for particular cells. A printout of the spreadsheet should be 'pasted in'. Record details of any software used and program developed to analyse results.

10 A detailed discussion of the outcomes of an experiment is generally not likely to be obtained during the course of an experiment when the laboratory notebook is being used. However, there may be some conclusions that are available at the time and can be entered in the notebook.

11 List any references consulted. This may be for details of theory or where values for constants have been taken from.

Figure 1.1 is an illustration of what part of a page in a laboratory notebook might look like.

1.3 Report writing

The aim of a report of an experiment is to communicate to readers what the experiment was about, how it was carried out and the findings. A report should be easy to read and be complete but concise. The layout of a report can considerably affect its clarity and ease of reading. A structured layout, with headings, is thus generally used. The following are typical report sections and a form of structure. In some cases, it may be appropriate to combine some sections under a single heading.

1   *Title*
    The report needs a clear and concise title which indicates what the experiment was about. For example, a title might be: The determination of the acceleration due to gravity, using a simple pendulum.

2   *Abstract*
    Often, and in particular if the experiment is a long one, there might be an abstract giving an overview of the experiment and its findings. This would typically be about 40 to 100 words and clearly and concisely indicate what was done and the major findings. For the simple pendulum experiment we might have:

27 October 1995 <u>Determination of the acceleration due to gravity, using a simple pendulum.</u>

<u>Apparatus</u>
Simple pendulum, a small heavy spherical bob attached to the end of a length of string about 2 m long.
Retort stand base, rod, boss and clamp with two metal strips to be used as jaws for clamping the end
of the pendulum string. Stopwatch 0.2 s, Metre rule

<u>Preliminary experiment</u>
Pendulum initially set at full length. Rough measurement indicates a time per oscillation of about 3 s.
Pendulum then set at a length of about 20 cm. Rough measurement indicates a time per oscillation of less than 1 s.
The error in a measurement is likely to be of the order of ±0.2 s, due to variations in human reaction time, thus to give accuracy
better than ±0.2% the time measured needs to be at least 100 s. Hence times were measured for 100 oscillations and repeated 5 times,

<u>Theory</u>

$T = 2\pi\sqrt{\frac{L}{g}}$ $L = \left(\frac{g}{4\pi^2}\right)T^2$ Graph of L against T^2, straight line with gradient $\frac{g}{4\pi^2}$

According to textbooks, the equation is only valid for small amplitude oscillations, angle from vertical less than 5° so that sin q
approximates to q.
See Newtonian Mechanics by A.P. French (Nelson) p440; Mechanics, Vibrations and Waves by T.B. Akrill and C.J. Millar (Murray),
p.256. Also assumed that the bob approximates to a point mass.

<u>Results</u>
Amplitudes of oscillations restricted to less than 5°.
Length measured from bottom of clamp to centre of bob, accuracy about ±2 mm
Calculator used for mean, standard deviation s_{n-1} key, and standard error.

Length (cm)	Time for 100 swings (s)							T (s)	T^2 (s²)
	1	2	3	4	5	Mean (s)	std error (s)		
200	284.0	283.1	284.1	285.1	283.9	284.1	±0.3	2.841 ± 0.003	8.066 ± 0.012
180	270.3	268.7	270.0	268.0	269.0	269.2	±0.4	2.692 ± 0.004	7.247 ± 0.015
160	255.2	253.0	253.8	254.1	255.2	254.3	±0.4	2.543 ± 0.004	6.462 ± 0.014
140	236.4	237.2	237.0	235.8	237.6	236.8	±0.4	2.368 ± 0.004	5.607 ± 0.013

etc.

Figure 1.1 *An example of part of a sheet in a laboratory notebook*

An object falling freely under the action of gravity moves with a constant acceleration, termed the acceleration due to gravity. This report describes one method, involving the measurement of the periodic time of a simple pendulum, that can be used to determine the acceleration in a laboratory.

3   *Introduction*
This gives the background to the experimental investigation and the particular aims of the investigation. For example we might have the

following for the experiment to determine the acceleration due to gravity using a simple pendulum:

When bodies freely fall in a vacuum they do so with a constant acceleration which is independent of the mass of the body. This acceleration is termed the acceleration due to gravity. The acceleration occurs because of the gravitational force between the mass of the Earth and the mass of the falling body. The force depends on the distance between the centres of the Earth and the body. However, for bodies close to the surface of the Earth and only falling a small distance, the force, and hence the acceleration, can be regarded as constant over the distance fallen. The aim of this experiment is to determine the acceleration due to gravity in the laboratory by measurements of the periodic time of a simple pendulum executing small amplitude oscillations.

4 *Method*

This section is devoted to a clear statement of how the experiment was carried out and the materials and instruments used. Any problems encountered should be given. Diagrams can be used to make the explanation clearer. Thus for the simple pendulum experiment we might have:

A small, but heavy, ball was suspended from a fixed point by a string. The resulting simple pendulum was allowed to swing back and forth to give small amplitude oscillations, the maximum angle θ of the string from the vertical being kept to less than 5° (Figure 1.2). A stopwatch was used to determine the time taken for 100 complete oscillations for a number of different lengths L of the pendulum. The reason for choosing 100 oscillations was to reduce the effects on the time measured of variable time lags between the moment the eye sends a message to the brain and the moment the hand operates the stopwatch. The length was measured, using a metre rule, from the point of support to the centre of the pendulum bob.

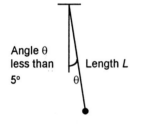

Figure 1.2 *Simple pendulum*

5 *Theory*

A statement of any theory or equation relevant to the investigation. For the simple pendulum experiment there might be:

The periodic time T of a simple pendulum depends on the length L of the pendulum and the acceleration due to gravity g, being given by:

$$T = 2\pi \sqrt{\frac{L}{g}}$$

The derivation of this equation[1,2] makes the approximation of sin θ to θ. For θ greater than 5°, such an approximation will result in errors greater than about 0.05%. For example, with θ = 24° the error is about 1%. The assumption is also made that the pendulum bob behaves as a point mass on the end of a massless string, otherwise the equation for a compound pendulum would need to be used.

References:
1. A.P. French, Newtonian Mechanics, 1971 (Nelson) page 440.
2. T.B. Akrill and C.J. Millar, Mechanics, Vibrations and Waves, 1974 (Murray) page 256.

6 *Results*
Details of the data obtained are then presented. A table is a concise and clear way of presenting the data.

7 *Interpretation of the results*
This section deals with the analysis of the data. Where there is significant errors in the quantities plotted, then the points may be shown as lines through the points, the line extending either side of a point to the extent of the errors (see Chapter 3). The linearity, or otherwise, of the resulting graph line is then discussed and deductions made from the graph. For example, in the case of the simple pendulum we might have:

The equation for the periodic time can be written as:

$$T^2 = \left(\frac{4\pi^2}{g}\right)L$$

Hence a graph of T^2 against L will be a straight line with a gradient of $(4\pi^2/g)$. Figure 1.3 shows the graph with the best line drawn through the points. The error in the gradient of the line was obtained by drawing the lines with the highest and lowest gradients feasible through the points.

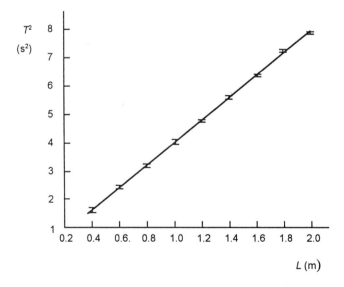

Figure 1.3 *Graph for the simple pendulum experiment*

8 *Conclusions*
This section refers back to the purpose of the experiment and indicates how far the experiment has gone to meet that purpose. Thus in the case of the simple pendulum experiment, the value of the acceleration due to gravity obtained in the experiment, together with its accuracy, is quoted. For comparison, the value obtained from tables might also be quoted and any difference commented on. For example you might have:

> Value obtained in this experiment $g = 9.79 \pm 0.03$ m/s^2
> Acceleration due to gravity at Teddington $g = 9.811\ 8118$ m/s^2
> (Kaye and Laby, Tables of Physical and Chemical Constants, 15th Ed. (Longman) page 160).

There might then be a comment about the result in relation to the quoted value. Though the locality differs and there could be some difference in g, within the limits of the experimental error of the experiment there is no significant difference.

1.3.1 Writing style

The following are some general points concerning the writing style to be used in writing reports of experiments:

1. In writing a report the normal style is to write in what is termed the third person. This means writing in the form:

    The time for 100 oscillations was measured for lengths of 0.40, 0.60, 0.80 and 1.0 m.

    rather than in the first person as:

    I measured the time for 100 oscillations for lengths of 0.40, 0.60, 0.80 and 1.0 m.

2. Reports should generally be written for the most part in the past tense, i.e. a statement of what was done and not as though it had been written while you were doing the experiment. Thus there might be:

    Measurements were made of the periodic time at a number of lengths.

    However, in the interpretation of the results we might have:

    The graph indicates that the square of the periodic time is proportional to the length.

3. In general, avoid long complicated sentences, short sentences tend to aid clarity. For example, consider the following two versions:

The derivation of this equation (see the references: A.P. French, Newtonian Mechanics, 1971 (Nelson) page 440; T.B. Akrill and C.J. Millar, Mechanics, Vibrations and Waves, 1974 (Murray) page 256) makes the approximation of sin θ to θ; however, for θ greater than 5°, such an approximation will result in errors greater than about 0.05%, e.g. with θ = 24° the error is about 1%, and for this reason the angle was restricted to less than 5°.

The derivation of this equation[1,2] makes the approximation of sin θ to θ. For θ greater than 5°, such an approximation will result in errors greater than about 0.05%. For example, with θ = 24° the error is about 1%. For this reason the angle was restricted to less than 5°.

References:
1. A.P. French, Newtonian Mechanics, 1971 (Nelson) page 440.
2. T.B. Akrill and C.J. Millar, Mechanics, Vibrations and Waves, 1974 (Murray) page 256.

The second version is easier to read and comprehend than the first one. Note one of the ways used to refer to references.

The following are some key points about sentence construction:

1. A sentence is a group of words which makes sense in itself, containing one item of information to which various subsidiary ideas may be attached.
2. A sentence, particularly if it contains technical information, should not be so long that the reader is unable to assimilate the data. Typically, such a sentence should be about 16 to 20 words with a maximum of about 40.
3. Variety in sentence length within a passage can aid the reader.
4. The main unit of a sentence should convey the main idea and must be readily identifiable by the reader.

4. Pages of text without breaks or spaces are overwhelming and difficult to read. Break up long pieces of prose by the use of paragraphs. A paragraph is formed from a series of ideas united by a theme.

5. Avoid roundabout ways of saying things. Keep things simple. For example, the second of the following versions is preferable to the first:

The calculation of the periodic time was made on the basis of using a mean value obtained from taking five measurements of the time.

The periodic time was calculated using the mean value from five measurements.

6. When abbreviations or symbols are first used, they should be explained. Thus we might have:

The periodic time of a simple pendulum depends on its length and the acceleration due to gravity, being given by:

$$T = 2\pi \sqrt{\frac{L}{g}}$$

where L is the length of the pendulum and g the acceleration due to gravity.

or alternatively, and more concisely:

The periodic time T of a simple pendulum depends on the length L of the pendulum and the acceleration due to gravity g, being given by:

$$T = 2\pi \sqrt{\frac{L}{g}}$$

7 After writing the report, review what you have written and see whether it makes 'sense' and is coherent and readable.

2 Experimental data

This chapter is about the presentation of the data obtained from experiments with correct units, using 'powers of ten' notation, and quoting data to a particular number of significant figures.

2.1 Units The *International System (SI)* of units has seven base units, these being:

Length	metre	m
Mass	kilogram	kg
Time	second	s
Electric current	ampere	A
Temperature	kelvin	K
Luminous intensity	candela	cd
Amount of substance	mole	mol

In addition there are two supplementary units, the radian and the steradian.

The SI units for other physical quantities are formed from the base units via the equation defining the quantity concerned. Thus, for example, volume is defined by the equation

volume = length cubed

The unit of volume is therefore that of unit of length unit cubed

unit of volume = unit of length cubed

Thus with the unit of length as the metre, the unit of volume is metre cubed, i.e. m^3.

Density is defined by the equation

$$\text{density} = \frac{\text{mass}}{\text{volume}}$$

Thus with the unit of density is the unit of mass divided by the unit of volume

$$\text{unit of density} = \frac{\text{unit of mass}}{\text{unit of volume}}$$

Thus, since the unit of mass is the kg and the unit of volume m^3, the unit of density is kg/m^3.

Velocity is defined, for motion in a straight line, by the equation

$$\text{velocity} = \frac{\text{change in distance covered}}{\text{time taken}}$$

Thus the unit of velocity is

$$\text{unit of velocity} = \frac{\text{unit of distance}}{\text{unit of time}}$$

Since the unit of distance, i.e. length, is the metre and the unit of time the second, then the unit of velocity is metres/second, i.e. m/s.
Acceleration is defined by the equation

$$\text{acceleration} = \frac{\text{change in velocity}}{\text{time taken}}$$

Thus the unit of acceleration is

$$\text{unit of acceleration} = \frac{\text{unit of velocity}}{\text{unit of time}}$$

Since the unit of velocity is metres per second and that of time the second, the unit of acceleration is

$$\text{unit of acceleration} = \frac{m/s}{s} = \frac{m}{s \times s} = m/s^2$$

Some of the derived units are given special names. Thus, for example, the unit of force is defined by the equation

force = mass × acceleration

and is thus

unit of force = unit of mass × unit of acceleration

and is thus kg m/s^2 or kg m s^{-2}. This unit is given the name newton (N). Thus 1 N is 1 kg m/s^2. The unit of pressure is given by the defining equation

$$\text{pressure} = \frac{\text{force}}{\text{area}}$$

and is thus

$$\text{unit of pressure} = \frac{\text{unit of force}}{\text{unit of area}}$$

Hence the derived unit of pressure is N/m^2. This unit is given the name pascal (Pa). Thus 1 Pa = 1 N/m^2. Table 2.1 lists examples of such derived units.

Table 2.1 *SI derived units with special names*

Quantity	Unit name	Symbol	Expressed in terms of:	
			other units	SI base units
Frequency	hertz	Hz		s^{-1}
Force	newton	N		$m\ kg\ s^{-2}$
Pressure and stress	pascal	Pa	N/m^2	$kg\ m^{-1}\ s^{-2}$
Energy, work, heat	joule	J	N m	$kg\ m^2\ s^{-2}$
Power	watt	W	J/s	$kg\ m^2\ s^{-3}$
Charge	coulomb	C		A s
Voltage	volt	V	W/A	$kg\ m^2\ A^{-1}\ s^{-3}$
Capacitance	farad	F	C/V	$s^4\ A^2\ kg^{-1}\ m^{-2}$
Resistance	ohm	Ω	V/A	$m^2\ kg\ A^{-2}\ s^{-3}$
Conductance	siemen	S	A/V	$s^3\ A^2\ kg^{-1}\ m^{-2}$
Magnetic flux	weber	Wb	V s	$m^2\ kg\ A^{-1}\ s^{-2}$
Magnetic flux density	tesla	T	Wb/m^2	$kg\ A^{-1}\ s^{-2}$
Inductance	henry	H	Wb/A	$m^2\ kg\ s^{-2}\ A^{-2}$

Certain quantities are defined as the ratio of two comparable quantities. Thus, for example, strain is defined as change in length/length. It thus is expressed as a pure number with no units because the derived unit would be m/m.

Note that sin θ, cos θ, tan θ, etc. are trigonometric ratios, i.e. θ is a ratio of two sides of a triangle. Thus θ has no units. The radian unit is in fact a non-unit since it represents a ratio. Thus, for sin ωt, when *t* has the unit of s, then ω must have the unit of s^{-1}.

When we have an equation with $X = Y$, then we must have the sizes of X and Y the same. We can thus have 2 = 2 or 5 = 5. But, when the quantities each side of the equals sign represent measurements then we must have the unit of X the same as the unit of Y. We can thus have X in m/s and Y in m/s, but not X in m/s and Y in m/s². This is the principle we have been using above to obtain the units of quantities. Now consider the application in more complex equations:

$$s = ut + \tfrac{1}{2}at^2$$

For the 'units to balance' we must have the unit of *s* to be the same as the unit of $(ut + \tfrac{1}{2}at^2)$. This means the unit of *s* must be the same as the unit of *ut* and also that of $\tfrac{1}{2}at^2$. Thus if the unit of *s* is metres, we must have:

unit of *ut* = m

and so, if the unit of t is seconds:

$$\text{unit of } u = \frac{m}{\text{unit of } t} = \frac{m}{s}$$

We also have:

unit of at^2 = m

and so:

$$\text{unit of } a = \frac{m}{\text{unit of } t^2} = \frac{m}{s^2}$$

Example

Determine the unit of the tensile modulus E when it is defined by the following equation:

$$E = \frac{\text{stress}}{\text{strain}}$$

Stress has the unit of Pa and strain is a ratio with no units.

The unit of E is thus given by:

$$\text{unit of } E = \frac{\text{unit of stress}}{\text{unit of strain}} = \text{Pa}$$

Example

In an experiment to determine the acceleration due to gravity using a simple pendulum, the following equation was used:

$$T^2 = \left(\frac{4\pi^2}{g}\right) L$$

The units used in the experiment for T were seconds and L centimetres. What would be the unit of g?

Rearranging the equation:

$$g = \frac{4\pi^2 L}{T^2}$$

and so:

$$\text{unit of } g = \frac{\text{unit of } L}{\text{unit of } T^2} = \frac{cm}{s^2}$$

2.1.1 Powers of ten notation

The term *scientific notation* or *standard notation* is often used to express large and small numbers as the product of two factors, one of them being a multiple of ten. For example, a voltage of 1500 V can be expressed as 1.5×10^3 V and a current of 0.0020 A as 2.0×10^{-3} A. The number 3 or -3 is termed the *exponent or power*. To write a number in powers we have to consider what power of ten number is used to multiply or divide it. Thus, for the voltage:

$$1500 = 1.5 \times 1000 = 1.5 \times 10^3$$

and for the current:

$$0.0020 = \frac{2.0}{1000} = 2.0 \times 10^{-3}$$

When numbers in scientific notation are multiplied, the exponents are added. Thus:

$$10^p \times 10^q = 10^{p+q}$$

Thus, for example:

$$10^2 \times 10^3 = 100 \times 1000 = 10^5$$

When there is division, the exponents are subtracted. Thus:

$$\frac{10^p}{10^q} = 10^{p-q}$$

Thus, for example:

$$\frac{10^5}{10^2} = \frac{100\,000}{100} = 1000 = 10^3$$

When we are concerned with the addition or subtraction of numbers in scientific notation, the numbers should be expressed using the same exponent. Then the factor of 10 to this exponent can be extracted from the addition or subtraction and the addition or subtraction operation carried out. For example:

$$4.502 \times 10^3 + 2.31 \times 10^2 = 45.02 \times 10^2 + 2.31 \times 10^2$$

$$= (45.02 + 2.31) \times 10^2$$

$$= 47.33 \times 10^2$$

When using a calculator to enter numbers in scientific notation the procedure is:

1. Key in the number.
2. Press the EXP key, then the value of the exponent and then, if the exponent is negative, the +/− key.

Thus, for example, to enter 4.05×10^5, key in 4.05, then press the EXP key, then 5. To enter 2.31×10^{-3}, key in 2.31, then press the EXP key, then 3 and finally the +/− key.

Example

Express the force 231 000 N in powers of ten.

The force can be considered to be $2.31 \times 100\ 000$ N and so written as 2.31×10^5 N.

2.1.2 Unit prefixes

Standard prefixes are used for multiples and submultiples of units, the SI preferred ones being multiples of 1000, i.e. 10^3, or division by multiples of 1000. Table 2.2 shows the standard prefixes.

Thus, for example, 1000 N can be written as 1 kN, 1 000 000 Pa as 1 MPa, 1 000 000 000 Pa as 1 GPa, 0.001 m as 1 mm, and 0.000 001 A as 1 μA.

Example

A capacitor is found to have a capacitance of 8.0×10^{-11} F. Express the capacitance in pF.

1 pF = 10^{-12} F. Hence, since:

$$8.0 \times 10^{-11} = 80 \times 10^{-12}$$

then the capacitance can be written as 80 pF.

Example

The tensile modulus for steel is about 210 GPa. Express this value in Pa without the unit prefix.

1 GPa = 10^9 Pa. Hence 210 GPa = 210×10^9 Pa.

Table 2.2 *Standard unit prefixes*

Multiplication factor	Prefix	
1 000 000 000 000 000 000 000 000 = 10^{24}	yotta	Y
1 000 000 000 000 000 000 000 = 10^{21}	zetta	Z
1 000 000 000 000 000 000 = 10^{18}	exa	E
1 000 000 000 000 000 = 10^{15}	peta	P
1 000 000 000 000 = 10^{12}	tera	T
1 000 000 000 = 10^{9}	giga	G
1 000 000 = 10^{6}	mega	M
1 000 = 10^{3}	kilo	k
100 = 10^{2}	hecto	h
10 = 10	deca	da
0.1 = 10^{-1}	deci	d
0.01 = 10^{-2}	centi	c
0.001 = 10^{-3}	milli	m
0.000 001 = 10^{-6}	micro	μ
0.000 000 001 = 10^{-9}	nano	n
0.000 000 000 001 = 10^{-12}	pico	p
0.000 000 000 000 001 = 10^{-15}	femto	f
0.000 000 000 000 000 001 = 10^{-18}	atto	a
0.000 000 000 000 000 000 001 = 10^{-21}	zepto	z
0.000 000 000 000 000 000 000 001 = 10^{-24}	yocto	y

Example

Determine the unit of pressure p when it is given by the equation $p = F/A$ and F has the unit of kN and A that of mm².

What we have is:

$$\text{unit of pressure} = \frac{\text{unit of } F}{\text{unit of } A} = \frac{\text{kN}}{\text{mm}^2} = \text{kN/mm}^2$$

We can write this unit in another way:

$$\text{unit of pressure} = \frac{1000 \times \text{N}}{0.001 \text{ m} \times 0.001 \text{ m}} = 10^9 \text{ N/m}^2 = \text{GPa}$$

2.1.3 Non-SI units

Other units which the reader may come across are fps (foot-pound-second) units which still are often used in the USA. On that system the unit of length is the foot (ft), with 1 ft = 0.3048 m. The unit of mass is the pound (lb), with 1 lb = 0.4536 kg. The unit of time is the second, the same as the SI system. With this system the derived unit of force, which is given a special name, is the poundal (pdl), with 1 pdl = 0.1383 N. However, a more common unit of force is the pound force (lbf). This is the gravitational force acting on a mass of 1 lb and consequently, since the standard value of the acceleration due to gravity is 32.174 ft/s², then

1 lbf = 32.174 pdl = 4.448 N

The similar unit the kilogram force (kgf) is sometimes used. This is the gravitational force acting on a mass of 1 kg and consequently, since the standard value of the acceleration due to gravity is 9.806 65 m/s², then

1 kgf = 9.806 65 N

A unit often used for pressure or stress in the USA is the psi, or pound force per square inch. 1 psi = 6.895×10^3 Pa.

2.1.4 Data in tables

When data is tabulated, the unit of measurement for data should be included with the label for a column or row of data. One way of doing this is to write the label for, say, a column of current values as Current (A). However, there is an alternative way and that is to write it as Current/A. This is based on the argument that tables only contain numbers so we must cancel out the units of the measured values by dividing by the unit. For example, if we have a reading of 0.020 A we can make this into just a number by dividing by the unit A and so have: Current/A 0.020. Alternatively we might write this as Current/mA 20, or perhaps Current/(10^{-3}A) 20.

When data in a table is expressed in powers of ten, the readability of a table is improved if the multiplying power of ten is included in the label of a column or row. Thus we might have Current ($\times 10^{-4}$ A) or Current/10^{-4} A. Such a label would indicate that all the current values in the column or row with that label have their values in A multiplied by 10^{-4}.

Example

Present the following current data in a table row: 1.01×10^{-2} A, 1.22×10^{-2} A, 1.41×10^{-2} A, 1.63×10^{-2} A, 1.82×10^{-2} A.

The following are alternative ways that might be used with such data:

Current (10^{-2} A)	1.01	1.22	1.41	1.63	1.82
Current/10^{-2} A	1.01	1.22	1.41	1.63	1.82
Current (mA)	10.1	12.2	14.1	16.3	1.82
Current/mA	10.1	12.2	14.1	16.3	18.2

2.2 Significant figures

All measurements are subject to inaccuracy. This means that it is never possible to quote an exact value for a quantity. Thus if the mass of some object lies between 11.95 g and 12.05 g we write it as 12.0 g. This indicates that the mass is certainly not 11.9 g or 12.1 g but nearer 12.0 g than either of these figures. As a working rule we tend to assume that a number has a *significance* equal to a single unit in the last figure given. Thus if we quote a number as 12.0 then we assume that the 0 is significant and that it means that the number is known as accurately as this figure. Thus it cannot be 12.1 or 11.9 but lies between these two values. The number is said to be known to three significant figures. However, if we quoted the number as 12, we are assuming that the 2 is significant and that the number is only known to the accuracy of that figure. Thus the number could be 12.1 or 11.9 since all we are saying is that the number must lie between 11 and 13. The number in this case has only been quoted to two significant figures. The number of significant figures is the number of figures which is accurately known.

If we have a number such as 0.0014 then the number of significant figures is 3 since we only include the number of figures between the first non-zero figure and the last figure. This becomes more obvious if we write the number in scientific notation as 1.04×10^{-3}. If we have a number written as 104 000 then we have to assume that it is written to six significant figures, the last 0 being significant. If we only wanted three significant figures then we should write the number as 1.04×10^5.

Often when using a calculator, the answer given will contain more digits than the numbers fed into it. For example, my calculator gives:

$2.0 \div 1.3 = 1.5384615$

The two numbers we started with are only known to two significant figures, yet the calculator has given an answer which implies eight significant figures. If we used a computer we might obtain an answer which implied perhaps 100 significant figures. All these extra figures are worthless. When multiplying or dividing two numbers, the result should only be given to the same number of significant figures as the number with the least number of significant figures. Thus the answer to the above division should just be quoted as 1.5. The figures could be measurements where the true values were 2.03 and 1.31, giving 1.5496183. There is thus uncertainty about the second figure after the decimal point. The accuracy to which the two numbers are known is such that we cannot claim any higher accuracy than implied by stating 1.5 as the result of the division.

When adding or subtracting numbers, the result should only be given to the same number of decimal places as the number in the calculation with the least number of decimal places. For example, 1.23 + 14.5 should be quoted as having the answer 15.7. This is because the 14.5 number is only known to one decimal place. It could, for example, be 14.52 or perhaps 14.54 and so when adding this number there is uncertainty about the second figure after the decimal point.

Some equations involve factors that are pure numbers, e.g. $A = \pi r^2$ with π being a pure number. For such numbers there is no inaccuracy and thus their values do not restrict the number of significant figures in any calculation.

When the result of a calculation produces a number which has more figures than are significant, we need to reduce it to the required number of significant figures. This process is termed *rounding*. For example, if we have 2.05 divided by 1.30, then using a calculator we obtain 1.5769231. We need to reduce this to three significant figures. This is done by considering the fourth figure, i.e. the 6. If that figure is 5 or greater, the third figure is rounded up. If that figure is less than 5, it is rounded down. In this case the figure of 6 is greater than 5 and so we round up and write the result as 1.58.

In any calculation which involves a number of arithmetic steps, do not round numbers until all the calculations have been completed. The rounding process carried out at each stage can considerably affect the number emerging as the final answer.

Example

In an experiment involving weighing a number of items the results obtained were 1.4134 g, 5.156 g and 131.5 g. Quote, to the appropriate number of significant figures, the result obtained by adding the weighings.

1.4134 + 5.156 + 131.5 = 138.0694. But one of the results is only quoted to one decimal place. Thus the answer should be quoted as 138.1 g, the second decimal place figure of 6 rounding the first figure up.

Example

The result of two measurements gave figures of 14.0 and 23.15. If we then have to determine a result by working out 14.0 divided by 23.15, what is the result to the appropriate number of significant figures?

14.0 ÷ 23.15 = 0.6047516. But the result with the least number of significant figures has just three. Hence the result to three significant figures is 0.605, the third figure having been rounded up because the fourth figure is 7.

Example

In an experiment to determine the tensile modulus of a length of steel wire, the following measurements were made:

Load applied $W = 5.0$ kg
Diameter of wire $d = 0.51$ mm
Extension of the wire $e = 1.84$ mm
Length of the wire $L = 1.541$ m

The equation used to calculate the modulus is:

$$\text{tensile modulus} = \frac{WgL}{\frac{1}{4}\pi d^2 e}$$

g is the acceleration due to gravity and can be taken as being 9.812 m/s². Determine, to the appropriate number of significant figures, the tensile modulus.

Putting the data in the equation, with all values put in terms of kg, m and s:

$$\text{tensile modulus} = \frac{5.0 \times 9.812 \times 1.541}{\frac{1}{4}\pi(0.51 \times 10^{-3})^2 \times 1.84 \times 10^{-3}}$$

Using a calculator, the value obtained is 2.0113151×10^{11} Pa. The least number of significant figures of any data is 2. Thus the result should be quoted as 2.0×10^{11} Pa.

Example

In an experiment to determine the molar mass M of a volatile liquid, 0.188 g of the liquid are injected into a syringe which contained 18.4 cm³ of air at a temperature of 56°C. The result is that the volume of gas in the syringe becomes 54.6 cm³. The atmospheric pressure p is 1.013×10^5 Pa. Determine the molar mass, to the appropriate number of significant figures, using the equation:

$$pV = \frac{m}{M}RT$$

m is the mass of the injected liquid, V is the volume occupied by the resulting vapour, T is the temperature on the Kelvin scale and R has the value 8.314 J K⁻¹ mol⁻¹.

Rearranging the equation gives:

$$M = \frac{mRT}{pV}$$

The temperature on the kelvin scale is 56 + 273 = 329 K, there being no point in considering any decimal places since the measured temperature has none. With masses in kg and volumes in cubic metres, putting the numbers in the equation gives:

$$M = \frac{0.188 \times 10^{-3} \times 8.314 \times 329}{1.013 \times 10^5 \times (54.6 - 18.4) \times 10^{-6}}$$

Hence, using a calculator, we obtain M = 0.1402633 kg/mol. The data values with the least number of significant figures have three. Hence the result should be quoted as 0.140 kg/mol or 140 g/mol.

Problems

1. Write the following data in scientific notation in the units indicated:
   (a) 20 mV in V, (b) 15 cm³ in m³, (c) 230 μA in A, (d) 20 dm³ in m³, (e) 15 pF in F, (f) 210 GPa in Pa, (g) 1 MV in V.

2. Write the following data in the units indicated:
   (a) 1.2×10^3 V in kV, (b) 2.0×10^5 Pa in MPa, (c) 0.20 m³ in dm³, (d) 2.4×10^{-12} F in pF, (e) 0.003 A in mA, (f) 12×10^9 Hz in GHz.

3. What is the unit of η in the following equation if p is in Pa, r in m, Q in m³/s and L in m?

$$\eta = \frac{\pi p r^4}{8LQ}$$

4. What is the unit of s in the following equation if V is in V, I in A, m in kg and θ_1 and θ_2 in K?

$$s = \frac{VI}{m(t_2 - t_1)}$$

5. What is the unit of f in the following equation if T is in N, m in kg and L in m?

$$f = \frac{1}{2L}\sqrt{\frac{T}{m}}$$

6. Round the following numbers to two significant figures:
   (a) 12.91, (b) 0.214, (c) 0.01391, (d) 191.9, (e) 0.0013499

7. The mass of a beaker was measured using a balance as 25.6 g. Water with a mass of 6.02 g was added to the beaker. What is the total mass to the appropriate number of significant figures?

8 In a determination of the coefficient of linear expansion of a metal in the form of a rod, the length of the rod was measured at 20°C as 17.93 cm and at 96°C as 24.95 cm. Determine, to the appropriate number of significant figures, the value of the coefficient if it is given by the following equation:

$$\text{coefficient} = \frac{L_t - L_0}{L_0(t - t_0)}$$

L_t is the length at t°C and L_0 at t_0°C.

9 The e.m.f. E of a thermocouple is related to the temperature t by the equation:

$$E = at + bt^2$$

If the e.m.f. is in mV and t in °C, what must be the units of the constants a and b?

3 Graphs

In experimental work in science and engineering, graphs are used to:

1. Enable trends in, and relationships between, experimental data to be more easily seen than are possible by just looking at the numbers. They are very good visual aids.

2. Determine the value of some quantity, such as that given by the gradient or the intercept when two quantities have a relationship that can be put in the form of a straight line graph.

3. Derive an empirical relationship between two quantities.

4. Compare experimental results with a theoretical graph.

This chapter is about the plotting of graphs, the straight line graph and how relationships can be put into such a form, and the determination of information from graphs.

3.1 Plotting graphs

In plotting graphs it is necessary to consider what quantities to plot along which axis and the scales to be used. It is also necessary to consider what form of ruling the graph paper should have.

With regard to the choice of axes, a convention that is generally used is that the *independent variable*, i.e. the quantity whose value is selected by the experimenter, is plotted along the horizontal axis and the *dependent variable*, i.e. the quantity whose value is determined as a consequence of the choice made by the experimenter of the independent variable, is plotted along the vertical axis. Thus, for example, in an experiment to determine the acceleration due to gravity by the use of a simple pendulum, the experimenter might choose to vary the length L of the pendulum and measure the resulting periodic times T. The length is then the independent variable and the time the dependent variable. If a graph is to be plotted to show the relationship between L and T, then the graph axis would be as shown in Figure 3.1.

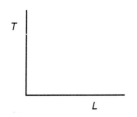

Figure 3.1 *Graph axes*

Each of the axes will have a scale. When selecting a scale, the following points should be taken into account:

1. The scale should be chosen so that the points to be plotted occupy the full range of the axes used for the graph. There is no point in having a graph with scales from 0 to 100 if all the data points have values between 0 and 50 (Figure 3.2(a)).

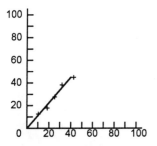

Figure 3.2 *(a) Badly chosen scales*

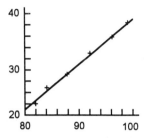

Figure 3.2 *(b) Graph with sensible choice of scales*

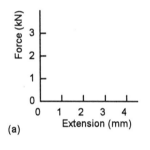

(a)

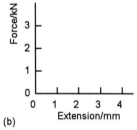

(b)

Figure 3.3 *Examples of labelling axes*

2   The scales should not start at zero if starting at zero produces an accumulation of points within a small area of the graph. Thus if all the points have values between 80 and 100, then a scale from 0 to 100 means all the points are concentrated in just the end zone of the scale. It is better, in this situation, to have a scale running from 80 to 100 (Figure 3.2(b)).

3   Scales should be chosen so that the location of the points between scale marks is made easy. Thus with a graph paper subdivided into large squares with each having 10 small squares, it is easy to locate a point of 0.2 if one large square corresponds to 1 but much more difficult if one large square corresponds to 3.

4   The axes should be labelled with the quantities they represent and their units. Figure 3.3 shows two ways that are used. In (b) the labels used for the axes are expressed in such a way that the measured quantities are divided by their units in order to give just numbers for the axes.

The scale selected may be *linear* with equal distances along the scale representing the same change in value, or *logarithmic* with equal distances on the axis representing a change by the same factor, e.g. 1 to 10 then 10 to 100. Special graph paper is available to make the use of logarithmic scales easier to use. Figure 3.4(a) shows the form that graph paper might take when both scales are linear, (b) when the independent scale is log and the dependent scale linear, (c) when both scales are log. The term *cycle* is used for each repetition of the rulings 1 to 10. Logarithmic graphs are discussed later in section 3.3.

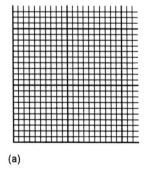

(a)

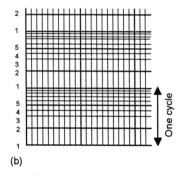

(b)

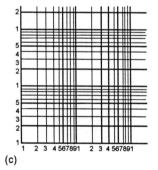

(c)

Figure 3.4 *Graph paper: (a) Linear-linear, (b) log-linear, (c) log-log*

3.1.1 Drawing graphs

Each experimental point should be clearly indicated on a graph. This may be by the use of crosses X, big dots ● or small dots with circles round them ⊙. A problem with big dots is that they give a loss of precision in the location of the point. This might, however, be required if there is some uncertainty associated with the point. Where there are noticeable errors associated with points, they may be represented as points with lines through them, the lengths of the lines giving an indication of the size of the error band associated with the measurement. Figure 3.5(a) shows the form this might take when there is an error in the dependent variable and the error in the independent variable is insignificant. Figure 3.5(b) shows the form it might take when there are significant errors in both the dependent and independent variables.

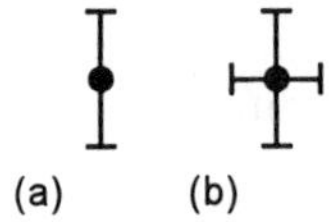

Figure 3.6 shows the error bars plotted for graphs resulting from the measurements of the extension of a spring when weights are added to it to stretch it. In (a) the vertical error bars indicate that the extension is measured to an accuracy of ±0.5 mm and the mass is known to an accuracy which is such that the error bar is not noticeable on the graph. In (b) the extension is known to an accuracy of ±0.5 mm and the mass to an accuracy of ±0.2 kg.

Figure 3.5 *Error bars*

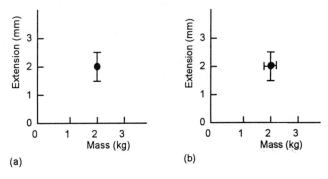

(a)   (b)

Figure 3.6 *Graphs with: (a) vertical error bars, (b) vertical and horizontal error bars*

When readings are being taken, they should generally be spread out reasonably evenly over the range of the quantities measured. Thus, in the case of the graph shown in Figure 3.7, the accuracy with which a line can be drawn through the points would be improved if readings were taken to provide points in the regions indicated. Where we need to locate some change, e.g. a transition from straight line to curve or from one straight line to another, then extra readings are worth taking in that region in order more accurately to locate the transition point.

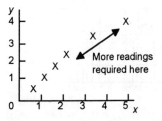

Figure 3.7 *Readings unevenly spread*

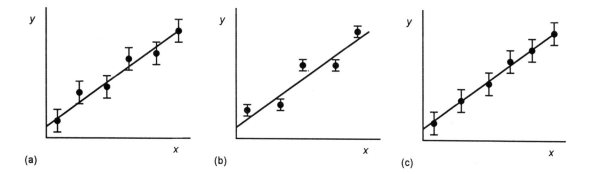

Figure 3.8 *(a) Smooth line through error bars, (b) possible underestimation of errors, (c) possible overestimation of errors*

Since, in most cases, graphs represent some smooth variation of one quantity with another, a straight line or a smooth curve should be drawn through the points plotted on a graph. Successive points should not be joined by short straight lines. The presence of error bars on a graph means that, if a smooth curve is expected, we might expect the results to be scattered about the smooth curve by amounts ranging up to the size of the error bar (Figure 3.8(a)). If the points deviate from the expected smooth curve by more than the error bands (Figure 3.8(b)) then it is possible that the errors may have been underestimated. A possible explanation might be that there is a permanent (systematic) error, e.g. an error due to a wrongly set zero on an instrument, which needs to be added to all the energy bands. If the readings all deviate from the expected smooth curve by less than the error bands (Figure 3.8(c)), then the errors might have been overestimated.

The *line of best fit* through a set of data points is the one drawn that has the data points scattered evenly above and below the line.

In some cases it might be that, within their error bands, all except one of the points lie on a smooth line. This suggests there might have been a mistake made when making that reading. This might have been the reading of the wrong scale on a multi-scale instrument. However, it might be that the point is not in error and that there is some abrupt change at that point. In cases like this it is worth checking the reading before discarding the point.

Example

The following data was obtained from an experiment in which the voltage was measured across a resistor at different currents. The reading error on each of the voltage readings was ±0.1 V and that on the current readings ±0.005 A. Plot the data as a graph.

Voltage (V)	0	0.5	1.0	1.5	2.0	2.5
Current (A)	0	0.12	0.19	0.30	0.40	0.48

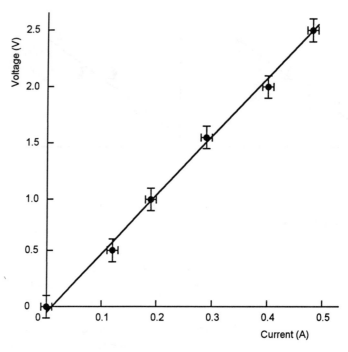

Figure 3.9 *Example*

Figure 3.9 shows the graph with the points plotted with their error bars and a smooth straight line plotted through them. Within the limits of their error bands, all the points lie on a straight line.

Example

The following data is for an experiment where the height of the free end of a cantilever above some fixed datum was measured when different loads were suspended from it. Plot the data as a graph of height against load. The heights are the average of a number of readings and all have an error of ±0.1 cm. The load has an accuracy of ±0.05 g.

Height (cm)	10.7	11.6	12.7	13.6
Load (g)	0	20	40	60

Figure 3.10 shows the graph. The scales of the graph have been chosen so that the height scale goes from 10 to 14 cm and the load from 0 to 60 g. This enables the points to be spread out over the graph and not all concentrated at one end. This would have occurred if the height scale had been from 0 to 14 cm. Only the error bar for the height is shown, since that for the load is not noticeable for the scale to which the graph has been drawn.

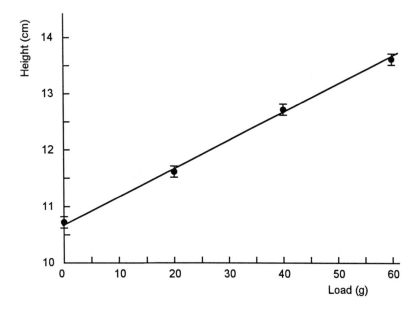

Figure 3.10 *Example*

Example

In an experiment, the resistance of a thermistor is measured at a number of temperatures and the following data obtained. The accuracy of the resistance measurements is ±0.1 kΩ and that of the temperature ±0.25°C. Plot a graph of the resistance against the temperature.

Resistance (kΩ)	3.8	2.4	1.6	1.1	0.7
Temperature (°C)	20	30	40	50	60

Figure 3.11 shows the graph. To spread the points out over the graph the scales have been chosen to go from 0 to 4 kΩ for the resistance and 20 to 60°C (extended to 70°C for the discussion in the next section on extrapolation) for the temperature. The error bars for the temperature are not noticeable with the scale chosen for the temperature. The result is a smooth curve through the points. If the error bands had been greater the straight line might have seemed feasible. However, within the error bands given, the result is a curve. See section 3.2.1 for details of how curves can be converted to straight line graphs.

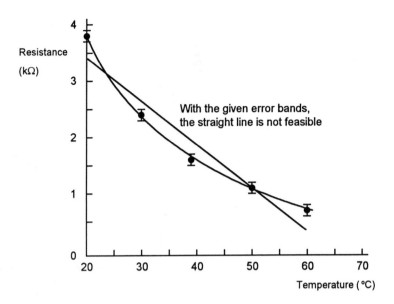

Figure 3.11 *Example*

3.1.2 Interpolation and extrapolation

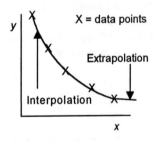

Figure 3.12 *Interpolation and extrapolation*

When the best line has been drawn through a set of data points, it is possible to use the line to determine values within the range of the data points. This is termed *interpolation* (Figure 3.12). For example, we can use Figure 3.11 to find the resistance of the thermistor at a temperature of 25°C. The value is about 3.0 kΩ.

If we want to find values that lie outside the range of the data points plotted then we use *extrapolation* (Figure 3.12). This means we extend the graph line to the region where no data was experimentally obtained and assume that the trend indicated by the graph continues in the same way. For example, we can use extrapolation with Figure 3.11 to estimate the resistance of the thermistor at a temperature of 70°C. The form of the graph curve would suggest a resistance of about 0.5 kΩ. Care must be exercised in using extrapolation; there may be some change which means that the graph line cannot just be extended.

3.2 Straight line graphs

The straight line graph is given with many relationships. Indeed we usually try to plot a graph in such a way that we can force it to become a straight line graph. In this section the form of equation describing a straight line graph is derived and discussed, with section 3.3 explaining how we can force relationships to give straight line graphs.

3.2.1 Equation for the straight line graph

Consider the straight line graph shown in Figure 3.13. The *gradient*, i.e. slope, of the graph is given by how much the line rises for a particular horizontal run and is thus:

$$\text{gradient } m = \frac{\text{rise}}{\text{run}} = \frac{y_2 - y_1}{x_2 - x_1}$$

Hence, we can write:

$$y_2 - y_1 = m(x_2 - x_1)$$

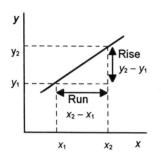

Figure 3.13 *Straight line graph*

The gradient is the same at all positions along a straight line. Thus we can take it to be as shown in Figure 3.14. For this straight line graph we have the line intercepting with the $x = 0$ axis at $y = c$. Then, if y and x are the set of values of some point on the line:

$$\text{gradient } m = \frac{\text{rise}}{\text{run}} = \frac{y - c}{x}$$

Thus:

$$y = mx + c$$

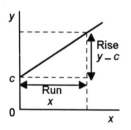

Figure 3.14 *Straight line graph with intercept on the y axis*

This is the general equation used to describe a straight line.

Example

Determine the equation of the straight line graph shown in Figure 3.15.

The gradient of the graph is:

$$m = \frac{\text{rise}}{\text{run}} = \frac{20 - 5}{4} = 3.75 \text{ m/s}$$

The graph has an intercept with the y axis of 5. Thus the equation is:

$$s = 3.75t + 5$$

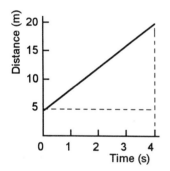

Figure 3.15 *Example*

where s is the distance in metres and t the time in seconds.

Example

Determine the equation of the straight line graph shown in Figure 3.16. Note that this is the graph plotted in an earlier example as Figure 3.10.

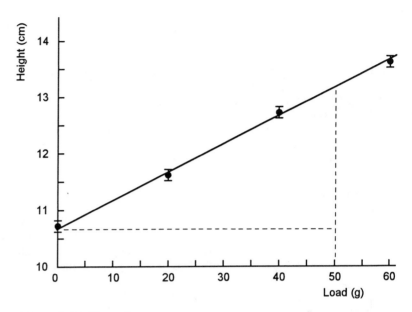

Figure 3.16 *Example*

The gradient of the graph is:

$$\text{gradient} = \frac{\text{rise}}{\text{run}} = \frac{2.5}{50} = 0.083 \text{ cm/g}$$

The graph has the intercept of about 10.7 with the y axis. Hence the equation is:

$$h = 0.083W + 10.7$$

where h is the height in centimetres and W the load in grams.

3.2.2 Uncertainties in gradient and intercept

Since there is uncertainty in the measurements used for drawing graphs, so there will be some uncertainties in the values estimated for the gradient and the intercept. For a graph drawn with error bars we can obtain an estimate of these uncertainties by considering the maximum and minimum gradient lines that can be drawn through the data.

Consider the graph shown in Figure 3.17. The solid line shows the best line that can be drawn through the error bars, the points being evenly scattered on either side of the line. The lines to give the maximum and minimum gradients are drawn so that they pass through all the error bars but tend towards the extremes of the bars. We can thus obtain the extreme values of the gradient and so obtain their differences from the best line value. The result can then be expressed as:

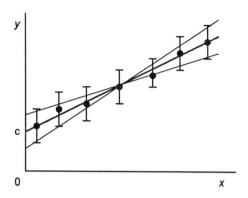

Figure 3.17 *Uncertainty estimation*

gradient = (gradient of best line) ± (difference from best line)

Likewise we can consider the maximum and minimum intercept values given by these lines and so obtain a measure of the uncertainty associated with the intercept value quoted.

3.3 Linearising equations

In many experiments we know the equation describing the relationship between two quantities and are attempting to find some constants in that equation. Often the relationship between the two quantities is such that it would not give a straight line graph. It is difficult to determine constants from a non-linear graph. For example, the periodic time t of a simple pendulum is related to the length of the pendulum by the equation:

$$T = 2\pi \sqrt{\frac{L}{g}}$$

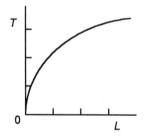

Figure 3.18 *Graph for simple pendulum*

A graph of T plotted against L does not give a straight line (Figure 3.18) and so it would be difficult to obtain a value of g from the graph. Thus we generally attempt to plot the results in such a way that a straight line graph is obtained and the constants can readily be determined.

Suppose we let $X = \sqrt{L}$. We can then write the above equation as:

$$T = \left(\frac{2\pi}{\sqrt{g}}\right) X$$

This is of the form $y = mx$ and so describes a straight line graph between T and X. Hence a graph of T against $\sqrt{L}$ is a straight line graph (Figure 3.19) with zero intercept and with a gradient of:

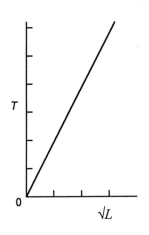

Figure 3.18 *Graph for simple pendulum*

$$\text{gradient} = \left(\frac{2\pi}{\sqrt{g}}\right)$$

An alternative way of obtaining a straight line graph would have been to write it as:

$$T^2 = \frac{4\pi^2}{g}L$$

If we then let $Y = T^2$ we have:

$$Y = \left(\frac{4\pi^2}{g}\right)L$$

This is of the form $y = mx$. Thus a graph of T^2 against L gives a straight line graph with zero intercept and a gradient of:

$$\text{gradient} = \frac{4\pi^2}{g}$$

See Chapter 1 and the example quoted there for such a graph.

The following are some examples of general equations and how they can be linearised:

1 $y = ax^2 + b$
 Let $X = x^2$ to give $y = aX + b$. A graph of y against x^2 is thus a straight line with a gradient of a and an intercept of b.

2 $y = (a/x) + b$
 Let $X = 1/x$ to give $y = aX + b$. A graph of y against $(1/x)$ is thus a straight line with a gradient of a and an intercept of b.

3 $y = ax^2 + bx$
 We can rearrange this equation to give $(y/x) = ax + b$. Let $Y = y/x$ to give $y = ax + b$. A graph of (y/x) against x is thus a straight line with gradient a and intercept b.

4 $y = ax^b$
 Taking logarithms gives $\lg y = \lg (ax^b) = \lg x^b + \lg a = b \lg x + \lg a$. If we let $Y = \lg y$ and $X = \lg x$ then we have $Y = bX + \lg a$. A graph of $\lg y$ against $\lg Y$ is thus straight line with a gradient of b and an intercept of $\lg a$.

5 $y = a\,e^{bx}$
 Taking logarithms to base e gives $\ln y = \ln (a\,e^{bx}) = \ln e^{bx} + \ln a$ which we can write as $\ln y = bx + \ln a$. If we let $Y = \ln y$ then $Y = bx + \ln a$. A graph of $\ln y$ against x is thus straight line with a gradient of b and an intercept of $\ln a$.

6 $y = a/\sin x$

If we let $X = 1/\sin x$ then we have $y = aX$. Thus a graph of y against $(1/\sin x)$ is a straight line with gradient a and zero intercept.

Example

The distance s travelled by a uniformly accelerating object after a time t is given by $s = ut + \frac{1}{2}at^2$, where u and a are constants. How can a straight line graph be obtained for the distance–time relationship?

Rearranging the equation:

$$\frac{s}{t} = \frac{1}{2}at + u$$

If we let $Y = s/t$ then we have $Y = \left(\frac{1}{2}a\right)t + u$. Thus a graph of s/t against t is a straight line with gradient of $(a/2)$ and an intercept of u.

Example

The current i in a circuit, when a capacitor is being charged, varies with time t according to the equation $i = I\,e^{t/RC}$, where I, R and C are constants. How can a straight line graph be obtained for the current–time relationship?

Taking logarithms to base e gives:

$$\ln i = \ln(I\,e^{t/RC}) = \left(\frac{1}{RC}\right)t + \ln I$$

Thus if we let $Y = \ln i$ we have $Y = (1/RC)t + \ln I$. Thus a graph of $\ln i$ against t is a straight line graph with gradient $(1/RC)$ and intercept $\ln I$.

Example

The volume of a gas is measured at a number of pressures p and the following results obtained. Show that the relationship is of the form $V = ap^b$ and determine the values of the constants a and b.

V/m³	13.3	11.4	10.0	8.9	8.0
p/(10^5 Pa)	1.2	1.4	1.6	1.8	2.0

Taking logarithms of the equation gives $\lg V = b \lg p + \lg a$. If we let $Y = \lg V$ and $X = \lg p$ we have $Y = bX + \lg a$. Thus a graph of $\lg V$ against $\lg p$ would give a straight line with gradient b and intercept $\lg a$. Putting the data in this form gives:

V/m^3	13.3	11.4	10.0	8.9	8.0
lg (V/m^3)	1.12	1.06	1.00	0.95	0.90
$p/(10^5 \text{ Pa})$	1.2	1.4	1.6	1.8	2.0
lg (p/Pa)	5.08	5.15	5.20	5.26	5.30

Figure 3.19 shows the graph of lg V against lg p. The graph is straight line so the equation does describe the relationship. By extrapolation we can work out that the intercept must be about 6.20. The gradient is about -1. The minus sign occurs because ln V is decreasing as ln p increases. Thus lg $a = 6.20$ and so $a = 16 \times 10^5$ and $b = -1$. The equation is thus $v = 16 \times 10^5 p^{-1}$.

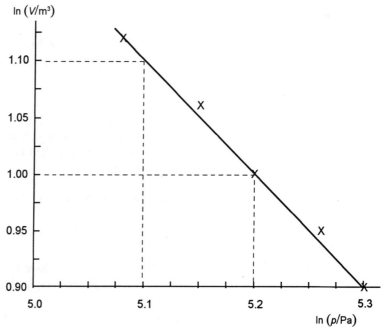

Figure 3.19 *Graph of lg V against lg p*

In the above example, the value of the gradient was obtained from an estimate of the gradient of the graph and the intercept by extrapolation. However, we could have obtained both these values, or certainly the intercept, by taking values of ln V and ln p from the straight line and substituting them in the original equation. Thus, if we take the estimation of the gradient as -1 from the graph, we can obtain a by substituting into the equation lg $V = b$ lg $p +$ lg a when we take values which lie on the straight line. Thus, taking the value lg $V = 1.00$ at lg $p = 5.2$ then $1.00 = -1 \times 5.2 +$ lg a. Hence, lg $a = 6.2$.

Note that the form used for expressing the measurements in the tables and on the graph is in the form of numbers. This is a particularly convenient way of writing units with log scales.

3.3.1 Log graphs

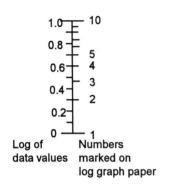

Figure 3.20 *A log scale*

When we have relationships involving powers, it is often the case that logarithms of the data are taken in order to give a straight line graph. The above example illustrates this. In order to plot that graph, the logarithms of each data point were taken. However, often a more convenient way to plot such a graph is to use logarithmic graph paper. With such graph paper, the conversion from a data point to its logarithm is done by the markings on the paper. Figure 3.20 illustrates this for one cycle.

We can have log–log or log–linear graph paper. Log–log graphs are used when the equation being plotted is of the form $y = ax^b$. Such an equation is used to describe the following data:

y	3.0	10	30	60	100
x	1	5	20	50	100

Figure 3.21 shows the data plotted on log–log graph paper. Notice how, in order to spread the data points out along the line, the values that have been measured are not equally spaced. If data, in the range $x = 1$ to $x = 100$, had been taken at, say, x values of 1, 20, 40, 60, 80 and 100 then all the graph points would have been concentrated in the top end of the graph line.

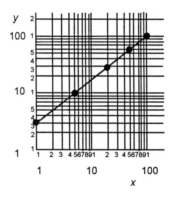

Figure 3.21 *Example of a log-log graph*

Log–linear graphs are used when the equation being plotted is of the form $y = a\,e^{bx}$. The following data is described by an equation of that form:

y	5.53	6.11	6.75	7.46	8.24
x	1	2	3	4	5

Figure 3.22 shows the data plotted on log–linear graph paper.

38 Experimental Methods

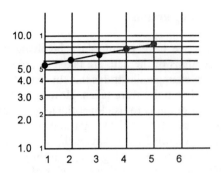

Figure 3.22 *Example of a log-linear graph*

3.3.2 Establishing relationships

Suppose an experiment is carried out and data obtained for two parameters; how can we establish, from the data, an equation relating the two? The first step is to plot the data, on a convenient set of scales, on linear graph paper. If the data points form a straight line then the equation is of the form $y = mx + c$ and the constants m and c can be evaluated. If the data does not give a straight line then, as a second step, we can try plotting the data on a log-log graph, either using log–log graph paper or converting the data values into log values. If the data gives a straight line then the equation is of the form $y = ax^b$ and the constants a and b can be evaluated. A third step is to try a log-linear graph, either using log–linear graph paper or converting data values into log values. If the data gives a straight line then the equation is of the form $y = a\,e^{bx}$ and a and b can be evaluated.

Example

The following data was obtained from an experiment in which the resistance of a thermistor was measured at a number of different temperatures. Determine an equation which relates the resistance with temperature.

Resistance (Ω)	16.3	6.25	2.66	1.24	0.63	0.34
Temperature (°C)	0	20	40	60	80	100

As a first step we could try a linear graph. However, since the resistance does not increase by equal amounts for equal changes in temperature, the data values would indicate that the graph would not be a linear one.

Suppose we try a log-log graph. The following table shows the logarithms of the values. Since it is likely that the temperature scale should be the kelvin scale, the values have been converted to that scale. This gets over the problem of the logarithm of 0.

Resistance	R/Ω	16.3	6.25	2.66	1.24	0.63	0.34
lg	(R/Ω)	1.21	0.80	0.42	0.09	−0.20	−0.47
Temperature	$t/°C$	0	20	40	60	80	100
Temperature	t/K	273	293	313	333	353	373
lg	(T/K)	2.44	2.47	2.50	2.52	2.55	2.57

Figure 3.23 shows the graph. While the results might indicate a straight line with the higher temperature results showing significant errors, it is likely that the graph is not a straight line so the relationship is not of the form $y = ax^b$.

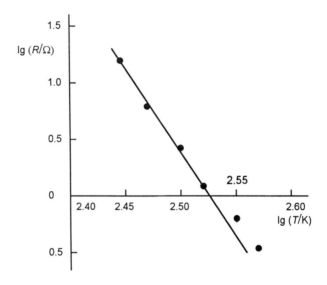

Figure 3.23 *Log-log graph*

As a third step we can try a log–linear graph. As before, we will use the temperature on the kelvin scale.

Resistance	R/Ω	16.3	6.25	2.66	1.24	0.63	0.34
ln	(R/Ω)	2.79	1.83	0.98	0.22	−0.46	−1.08
Temperature	$t/°C$	0	20	40	60	80	100
Temperature	t/K	273	293	313	333	353	373

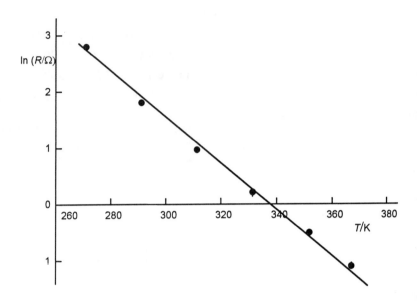

Figure 3.24 *Log-linear graph*

Figure 3.24 shows the result. The points do lie on a straight line. Thus the equation is of the form $R = a\, e^{bT}$. Since $\ln R = bT + \ln a$, the gradient of the graph gives b and the intercept with the y axis gives $\ln a$. The gradient is about -0.032 and the intercept, by extrapolation or calculation, about 11.5. This gives a value for a of about 98 700. Thus the equation is:

$$R = 98\,700\, e^{-0.032T}$$

3.4 Areas under graphs

In some situations it is necessary to estimate the area enclosed between a graph line and the axis of a graph. This might, for example, be an estimation of the work done when a spring, or other elastic member, is stretched through some distance. With a straight line graph, the area can easily be determined. Thus for Figure 3.25, the marked area is the area of a right-angled triangle and so is:

$$\text{area} = \tfrac{1}{2}Fx$$

Where the graph is non-linear, a simple method that can be used is to count the squares on the graph paper under the line and then multiply the number by the value corresponding to their area. Another way is to use the *mid-ordinate rule*. The area under the graph is divided into a number of equal width vertical strips (Figure 3.26). The area of a strip is obtained by multiplying the mid-ordinate of the strip by its width and the total area is

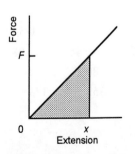

Figure 3.25 *Area under the graph line*

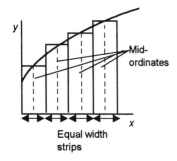

Figure 3.26 *The mid-ordinate rule*

then obtained by summing the contributions occurring from each of the strips:

area = Σ (mid-ordinate × strip width) = strip width × Σ (mid-ordinates)

Example

Use the mid-ordinate rule to determine the area under a velocity–time graph, between the times $t = 0$ and $t = 60$ s, obtained from the following data:

Velocity (m/s)	0	14	20	24	27	28	29
Time (s)	0	10	20	30	40	50	60

Figure 3.27 shows the graph with the area divided into six vertical strips, each of width 10 s. The mid-ordinate value for the first strip is about 9 m/s, for the second strip about 17 m/s, for the third strip about 22 m/s, for the fourth strip about 25.5 m/s, for the fifth strip about 27.5 m/s and for the final strip about 28.5 m/s. The total area is thus:

area = (9 + 17 + 22 + 25.5 + 27.5 + 28.5) × 10 = 1295 m

The accuracy would be improved if we used smaller width strips.

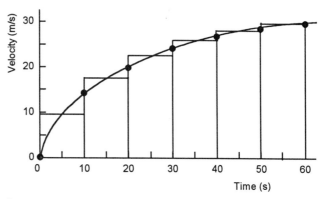

Figure 3.27 *Example*

Problems 1 The tensile strength σ of an alloy has been measured at a number of temperatures θ and the following results obtained. The tensile strengths were measured to an accuracy of ±0.1 kPa and the temperatures to ±1°C. The equation relating the tensile strength and temperature is expected to be of the form $\sigma = a\theta + b$. By plotting a suitable graph, verify that this is the case and obtain the values of the constants a and b.

σ (kPa)	83.7	80.1	77.7	75.3	72.7
θ (°C)	100	200	300	400	500

2   The rate of rotation n of the shaft of a motor is measured at various applied voltages V and the following results obtained. The rate of rotation was measured to an accuracy of ±10% and the voltages to ±0.5 V. The equation relating the rate of rotation and the voltage is expected to be of the form $n = aV + b$. By plotting a suitable graph verify that this is the case and determine the values of the constants a and b.

n (rev/s)	3	10	16	20	26	33
V (V)	40	80	120	160	200	240

3   The load W lifted by a machine is measured when different efforts E are applied and the following results obtained. Both quantities are measured to an accuracy of ±0.5 N. The equation relating the load and the effort is expected to be of the form $E = aW + b$. Verify, by plotting a suitable graph, that this is the case and determine the values of the constants a and b.

E (N)	18	27	32	43	51
W (N)	40	80	120	160	200

4   The variation of the reaction rate constant k with temperature T is described by the equation $k = A\,e^{-E/RT}$, where A and E are constant for a given reaction and R is the gas constant (8.31 J K^{-1} mol^{-1}). In an experimental investigation of the reaction:

$$2N_2O_5(g) \rightarrow 2N_2O_4(g) + O_2(g)$$

the following values were obtained for k at a number of temperatures. Determine, by drawing a suitable graph, the values of A and E.

k (s^{-1})	1.76×10^{-5}	1.00×10^{-4}	4.98×10^{-4}	2.03×10^{-3}
T (K)	293	307	318	333

5   The resistance R of a lamp is measured at a number of voltages V and the following data obtained. The resistances were measured to an accuracy of ±1 Ω and the voltages to ±0.5 V. Show, by plotting a suitable graph, that the equation relating the resistance and voltage is of the form $R = (a/V) + b$ and determine the values of a and b.

R (Ω)	70	62	59	56	55
V (V)	60	100	140	200	240

6 The periodic time T was measured for a simple pendulum with various lengths L and the following results obtained. Verify, by plotting a suitable graph, that the periodic time is related to the length by an equation of the form $T = aL^b$ and determine the values of a and b.

T (s)	1.41	1.55	1.67	1.79	1.90
L (m)	0.50	0.60	0.70	0.80	0.90

7 The rate of flow Q of water over a V-shaped notch weir was measured for different heights h of the water above the point of the V and the following data obtained. Verify, by plotting a suitable graph, that the rate of flow is related to the height by an equation of the form $Q = ah^b$ and determine the values of the constants a and b.

Q (m³/s)	0.13	0.26	0.46	2.12	1.07
h (m)	0.30	0.40	0.50	0.60	0.70

8 The voltage v across a component in an electrical circuit is measured at a number of times t and the following results obtained. Determine the equation relating v and t.

v (V)	3.75	1.38	0.51	0.19	0.07
t (s)	10	20	30	40	50

9 In a chemical experiment, 0.1 mol of propanone and 0.01 mol of hydrochloric acid in a total volume of 90 cm³ of water were placed in a flask at constant temperature. A clock was started when 0.0004 mol of iodine in 10 cm³ of water were added. Every 5 minutes, 10 cm³ samples were taken and neutralised by adding to excess aqueous sodium hydrogencarbonate. The samples were then titrated with 0.01 mol/dm³ sodium thiosulphate solution and the following results obtained. The volumes of sodium thiosulphate were determined to an accuracy of ±0.1 cm³ and the times to ±1 s. Determine, by drawing a suitable graph, the equation relating the volume of sodium thiosulphate and the time for which the reaction had proceeded.

Volume (cm³)	8	7.2	6.5	5.7	5.0	4.2
Time (min)	0	5	10	15	20	25

10 An experiment involves suspending a horizontal bar by two equal length parallel strings and measuring the periodic time T for horizontal oscillations, the arrangement being termed a biflar pendulum. The periodic time was measured for different horizontal separations s of the strings, the lengths being kept constant. The following are the results of such measurements. The periodic time measurements, the results of taking the mean of several readings, have an accuracy of ±0.03 s and

the separation an accuracy of ±0.1 cm. By plotting a suitable graph, determine the equation relating the periodic time and the separation.

T (s)	2.17	1.97	1.75	1.50	1.31
s (cm)	20.0	22.0	25.0	29.0	33.0

11 The magnification m produced by a thin convex lens of focal length f depends on the distance from the lens of the image v. The equation relating these quantities being:

$$m = \frac{v}{f} - 1$$

In an experiment an illuminated circular hole of diameter 10 mm was used for the object and the image produced on a screen on which circles of different diameter had been drawn. The distance of the object from the lens was varied and the image distance measured at which different magnifications were produced. The following results were obtained:

v (cm)	20.4	22.9	24.9	30.4	35.2
m	1.00	1.25	1.50	2.00	2.50

The accuracy of the image distance was ±0.3 mm and that of the magnification ±0.05. Determine, by the use of a suitable graph, the focal length of the lens.

12 The current i through a component in an electrical circuit was measured at a number of times t and the following results obtained. By drawing a suitable graph, determine the equation relating the current and time.

i (μA)	0	78	61	47	37	29	22
t (s)	0	5	10	15	20	25	30

13 In an experiment, the coefficient of friction μ was measured between a steel wheel and a steel surface for a range of velocities v of the wheel over the surface. The coefficient of friction data has an accuracy of ±0.01 and the velocities 0.5 m/s. The following results were obtained. By plotting a suitable graph determine the equation relating the coefficient of friction and the velocity.

μ	0.14	0.07	0.04	0.03	0.02
v (m/s)	10	20	30	40	50

14 The temperature θ of a hot solid in still air was measured as it cooled over a period of time t and the following results obtained. By plotting a suitable graph, determine the equation relating θ and t.

θ (°C)	50	37	28	20	15
t (min)	0	5	10	15	20

15 The following data was obtained for the velocity v of an object starting from rest at a time $t = 0$. Plot a graph of velocity against time and estimate the distance covered in 8 s from rest. The distance covered is the area under the graph between the $t = 0$ and $t = 8$ s ordinates.

v (mm/s)	0	8.8	12.6	16.3	21.4	22.5	21.6	17.3	12.7
t (s)	0	1	2	3	4	5	6	7	8

4 Experimental errors

This chapter is an introduction to experimental errors with a consideration of the importance of quoting them, their sources and types and how they can be combined. Chapter 5 takes the discussion of errors further with a consideration of the statistics involved.

4.1 Why estimate errors?

When a physical quantity is measured, the value obtained should not be expected to be exactly equal to the true value. For example, in a measurement of the acceleration due to gravity, the result calculated from a single measurement of the time taken for a ball to fall through a measured height might be 9.90 m/s². It is not likely that this will be the true value, indeed if the experiment is repeated it is extremely likely that a different value for the acceleration would be obtained. It is thus important to give some indication of how close to the true value a result might be expected to be. The term *accuracy* is used for the extent to which a result might depart from the true value. This can be done by including with the result an estimate of the error. *Error* is defined as being the difference between the result of a measurement and the true value (Figure 4.1):

$$\text{error} = \text{measured value} - \text{true value}$$

Thus the result of the measurement of the acceleration due to gravity g might be quoted as:

$$g = 9.90 \pm 0.15 \text{ m/s}^2$$

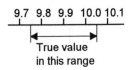

Figure 4.1 *Accuracy of measured value*

This means that we expect the acceleration due to gravity to be somewhere in the range 9.90 − 0.15 = 9.75 m/s² to 9.90 + 0.15 = 10.05 m/s² (Figure 4.1). The more accurate the measurement the smaller will be the error range. We cannot be certain that the value lies within this range but there is a certain probability that it will (see Chapter 5 for a discussion of the probability).

The quoting of the errors associated with a measurement are important if any meaningful conclusions are to be drawn from the result. For example, an earlier measurement of the acceleration due to gravity might have given a value of:

$$g = 9.80 \pm 0.05 \text{ m/s}^2$$

Does the later result indicate a change in the acceleration? Without quoting the errors the values of 9.80 and 9.90 m/s² might seem to indicate a change. However, when we consider the errors we have the later measurement giving a value expected to be between 9.75 and 10.05 m/s² and the earlier measurement a value between 9.75 and 9.85 m/s². Since these error ranges

Experimental errors 47

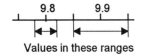

Values in these ranges

Figure 4.2 *Accuracy of measured values*

Values in these ranges

Figure 4.3 *Change in value*

overlap (Figure 4.2), the true value might not have changed and we have no evidence to suggest that it did.

However, suppose we had the two results:

Earlier measurement: $g = 9.80 \pm 0.02$ m/s²

Later measurement: $g = 9.90 \pm 0.05$ m/s²

The earlier measurement indicates a value in the range 9.78 to 9.82 m/s² while the later measurement indicates the range 9.85 to 9.95 m/s² (Figure 4.3). The ranges do not overlap and we are led to the conclusion that the acceleration due to gravity has changed.

Without quoting the error with an experimental result, it is not possible to judge the significance of a result and the result is really useless. Errors must be quoted with all experimental results.

4.1.1 Fractional and percentage errors

In some situations the error is specified in the form of the *fractional error*, where

$$\text{fractional error} = \frac{\text{error in quantit}}{\text{size of quantity}}$$

More frequently, however, the *percentage error* is quoted:

$$\text{percentage error} = \frac{\text{error in quantity}}{\text{size of quantity}} \times 10$$

Thus, for example, a speed quoted as 2.0 ± 0.2 m/s might have its error quoted as ±10%.

In the case of some instruments, the error is quoted as a percentage of the full-scale-reading that is possible with an instrument. Thus, for example, an ammeter used to measure a current using its 0 to 5 A range and having an error quoted as ±1% will give for all its readings on that range an error of ±0.05 A.

4.2 Sources of error

Common sources of error with measurements are:

1 *Instrument construction errors*
   These result from such causes as tolerances on the dimensions of components and the values of electrical components used in instruments and are inherent in the manufacture of an instrument. In addition there can be errors due to the accuracy with which the manufacturer of an instrument has calibrated it. For example, a cheap ruler might have divisions that are not equally spaced.

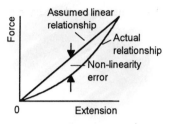

Figure 4.4 *Non-linearity error*

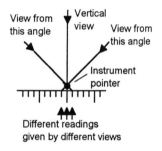

Figure 4.5 *Parallax error*

2 *Non-linearity errors*
   In the design of many instruments a linear relationship between two quantities is often assumed, e.g. a spring balance assumes a linear relationship between force and extension. This may be an approximation or may be restricted to a narrow range of values. Thus an instrument may have errors due to a component not having a perfectly linear relationship (Figure 4.4).

3 *Operating errors*
   These can occur for a variety of reasons and are often referred to as *human errors*. They can also include errors due to carelessness in reading a scale, perhaps reading the wrong scale and may occur in reading the position of a pointer on a scale. If the scale and the pointer are not in the same plane then the reading obtained depends on the angle at which the pointer is viewed against the scale (Figure 4.5). These are called *parallax errors*. To reduce the chance of such errors occurring, some instruments incorporate a mirror alongside the scale. Positioning the eye so that the pointer and its image are in line guarantees that the pointer is being viewed at the right angle. Digital instruments, where the reading is displayed as a series of numbers, avoids this problem of parallax. Errors may also occur due to the limited resolution of an instrument and the ability to read a scale. Such errors are termed *reading errors* (see section 4.2.1). Operating errors can also arise when an instrument has to be brought into contact with an object being measured, e.g. a micrometer, as a result of slightly different contact forces occurring.

4 *Environmental errors*
   Errors can arise as a result of environmental effects. For example, when making measurements with a steel rule, the temperature when the measurement is made might not be the same as that for which the rule was calibrated. Another example might be the presence of draughts affecting the readings given by a balance.

5 *Insertion errors*
   In some measurements the insertion of the instrument into the position to measure a quantity can affect its value. For example, inserting an ammeter into a circuit to measure the current can affect the value of the current due to the ammeter's own resistance (see section 4.2.2). Similarly, putting a cold thermometer into a hot liquid can cool the liquid and so change the temperature being measured.

6 *Unrepresentative samples*
   There are many situations where a sample of a material is taken and measurements made on it with the assumption being made that these are the properties of the material as a whole. For example, an engineer might carry out a tensile test of a sample of steel in order to forecast the properties of that steel when it is used in constructing a bridge. The

sample must be chosen in such a way that its properties are representative, otherwise errors will occur.

4.2.1 Reading errors

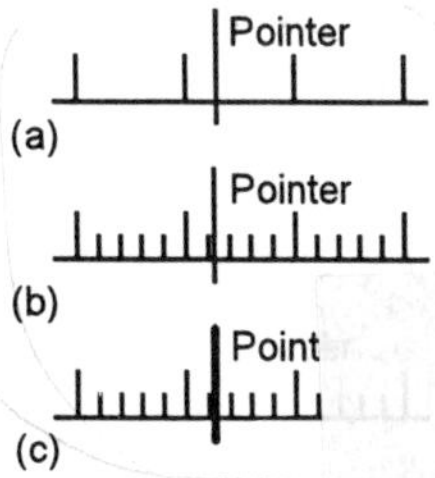

Figure 4.7 *Reading errors*

One particular form of operating error associated with the scale markings on an instrument is the *reading error*. When the pointer of an instrument falls between two scale markings there is some degree of uncertainty, i.e. possible error, as to what the reading should be quoted as. Consider the two scales shown in Figure 4.7. The scale in (a) has fewer graduations than that in (b). It is thus possible to read the position of the pointer in (b) with less uncertainty than that in (a). The magnitude of the reading error made will depend to some extent on the coarseness of the scale graduations. It also depends to some extent on the width of the pointer, as Figure 4.7(b) and (c) illustrate. The finer the pointer the more precise we can be as to its location on the scale.

Because no instrument has a scale which gives a reading to an infinitely fine resolution, there is always some uncertainty regarding a scale reading. Thus a reading should not be quoted as a precise number but some indication given of the uncertainty associated with it, i.e. the possible extent to which the reading could be in error. The worse the reading error could be is that the value indicated by a pointer is somewhere between two successive markings on the scale. In such circumstances the reading error can be stated as a value ± half the scale interval. For example, a rule might have scale markings every 1 mm. Thus when measuring a length using the rule, the result might be quoted as 23.4 ± 1 mm. However, it is often the case that we can be more certain about the reading and indicate a smaller error.

With digital displays there is no uncertainty regarding the value displayed but there is still an error associated with the reading. This is because the reading of the instrument goes up in jumps, a whole digit at a time. We cannot tell where between two successive digits the actual value really is. Thus the degree of uncertainty is ± the smallest digit.

4.2.2 Insertion errors

Figure 4.8 *Loading by a voltmeter*

When the act of making a measurement modifies the variable being measured, the term *loading* is used. As an illustration of such errors, consider a frequently encountered loading problem, that due to a voltmeter being used to measure the voltage across some resistor. Any electrical network may be considered to behave as a circuit containing a single source of e.m.f. in series with a resistance (this assumes we are considering a d.c. circuit) and thus we can represent a network by such an arrangement supplying a current through a load resistor (Figure 4.8). In the absence of the voltmeter the current through the load resistor, resistance R_L, is I and so the potential difference V_L across it is IR_L. If the rest of the circuit has a resistance R and the supplied e.m.f. is E, then:

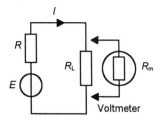

$$E = I(R + R_L)$$

and so the potential difference across the load is:

$$V_L = IR_L = E\left(\frac{R_L}{R + R_L}\right)$$

When the voltmeter, resistance R_m, is connected across the resistance, then we have a resistance R_m in parallel with R_L. For resistances in parallel, the equivalent resistance R_e is given by:

$$\frac{1}{R_e} = \frac{1}{R_L} + \frac{1}{R_m} = \frac{R_m + R_L}{R_L R_m}$$

The total resistance of the circuit will change and so the circuit current will change, becoming $E/(R + R_e)$. Thus the voltage V_m indicated by the voltmeter is:

$$V_m = E\left(\frac{R_e}{R + R_e}\right)$$

There will thus be an error, with V_m not equal to V_L, because the voltmeter does not have an infinite resistance. The error is $(V_m - V_L)$ and the fractional error is $(V_m - V_L)/V_L$. Hence:

$$\text{fractional error} = \frac{V_m - V_L}{V_L} = \frac{E\left(\frac{R_e}{R + R_e}\right) - E\left(\frac{R_L}{R + R_L}\right)}{E\left(\frac{R_L}{R + R_L}\right)}$$

$$= \frac{\left(\frac{R_e(R + R_L) - R_L(R + R_e)}{(R + R_e)(R + R_L)}\right)}{\left(\frac{R_L}{R + R_L}\right)}$$

$$= \frac{R(R_e - R_L)}{R_L(R + R_e)}$$

Since R_e will be less than R_L, the fractional error will be negative. The error will only be zero when R_e is equal to R_L and this will only occur when the voltmeter resistance is effectively infinite.

Example

A voltmeter with a resistance of 10 kΩ is placed in parallel with a load resistance of 1 kΩ which is in a circuit of total resistance 2 kΩ (Figure 4.9). What will be the fractional error in the measured voltage?

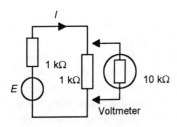

Figure 4.9 *Loading by a voltmeter*

The equivalent resistance when the voltmeter is in parallel with the load is given by

$$\frac{1}{R_e} = \frac{1}{10} + \frac{1}{1}$$

Thus $R_e = 0.91$ kΩ. Hence the fractional error in the voltage reading is:

$$\text{fractional error} = \frac{1(0.91 - 1)}{1(1 + 0.91)} = -0.047$$

4.3 Random and systematic errors

All errors, whatever their source, can be described as being either random or systematic. *Random errors* are ones which can vary in a random manner between successive readings of the same quantity. These may be due to personal fluctuations by the person making the measurements, e.g. varying reaction times in timing events, applying varying pressures when using a micrometer screw gauge, parallax errors, etc., or perhaps due to random electronic fluctuations (termed noise) in the instruments or circuits used, or perhaps varying frictional effects. *Systematic errors* are errors which do not vary from one reading to another. These may be due to some defect in the instrument such as a wrongly set zero so that it always gives a high or low reading, or perhaps incorrect calibration, or perhaps an instrument is temperature dependent and the measurement is made under conditions which differ from those for which it was calibrated, or there is a loading error. Random errors can be minimised by the use of statistical analysis, systematic errors require the use of a different instrument or measurement technique to establish them.

4.4 The mean value and its error

Random errors mean that sometimes the error will give a reading that is too high, sometimes a reading that is too low. The error can be reduced by repeated readings being taken and calculating the mean (or average) value. The *mean or average* $\bar{x}$ of a set of readings is given by

$$\bar{x} = \frac{x_1 + x_2 + \ldots x_n}{n}$$

where x_1 is the first reading, x_2 the second reading, ... x_n the nth reading. The more readings we take the more likely it will be that we can cancel out the random variations that occur between readings. The *true value* might thus be regarded as the value given by the mean of a very large number of readings.

Note that many pocket calculators have a built-in facility for determining mean values. With my Casio calculator the key for determining the mean is labelled as $\bar{x}$. The sequence of operations for calculating the mean from a set of measurements is:

1. Set the calculator to the statistics mode by, with my calculator, pressing SHIFT and MODE. The display then shows SD.

2. Key in the first reading. Then press the M+ key

3 Key in the second reading. Then press the M+ key.

4 Repeat these steps for each reading.

5 Finally press the $\bar{x}$ key. The mean is then displayed.

6 If you want to check the number of data points used to obtain the mean, press the *n* key.

Example

Five measurements of the time have been taken for 10 oscillations of a simple pendulum (Figure 4.10). As a result of variations in the reaction time and perception of when the pendulum has reached a particular point in its swing, random variations occur with the results: 20.1, 20.0, 20.2, 20.1, 20.1 s. Determine the mean value.

The mean value is obtained using the equation given above as:

$$\text{mean} = \frac{20.1 + 20.0 + 20.2 + 20.1 + 20.1}{5} = \frac{100.5}{5} = 20.1 \text{ s}$$

Figure 4.10 *Timing a simple pendulum*

4.4.1 Error of a reading

If we consider a single result, what is the likely error from the mean value? Some indication as to the degree of uncertainty is given from considering the spread of results obtained when repeated measurements are made. Consider the two following sets of readings:

20.1, 20.0, 20.2, 20.1, 20.1 and 19.5, 20.5, 19.7, 20.6, 20.2 s

Both the sets of readings have the same average of 20.1, but the second set of readings is more spread out than the first and thus shows more random fluctuations.

The *deviation* of any one reading from the mean is the difference between its value and the mean value. Table 4.1 shows the deviations for the two sets of results.

The second set of readings has greater deviations than the first set. It seems likely that if we had only considered one reading of the less spread out set of readings it would have had a greater chance of being closer to the mean value than any one reading in the more spread out set. The spread of the readings is thus taken as a measure of the certainty we can attach to any one reading being close to the mean value, the bigger the spread the greater the uncertainty. The spread of the readings is specified by a quantity termed the *standard deviation*.

Table 4.1 *Deviations from the mean*

1st set of readings		2nd set of readings	
Reading (s)	Deviation (s)	Reading (s)	Deviation (s)
20.1	0.0	19.5	−0.6
20.0	−0.1	20.5	+0.4
20.2	+0.1	19.7	−0.4
20.1	0.0	20.6	+0.5
20.1	0.0	20.2	+0.1

The standard deviation, symbol σ, is given by:

$$\text{standard deviation} = \sqrt{\frac{(d_1^2 + d_2^2 + \ldots d_n^2)}{n-1}}$$

where d_1 is the deviation of the first result from its average, d_2 the deviation of the second reading, ... d_n the deviation of the nth reading from the average. Note that sometimes the above equation is written with just n instead of $(n-1)$ on the bottom line. With just n it is assumed that the deviations are all from the true value, i.e. the mean when there are very large numbers of readings. With smaller numbers of readings, the deviations are taken from the mean value without assuming that it is necessarily the true value. To allow for this, $(n-1)$ is used. In fact, with more than a very few readings, the results using n and $(n-1)$ are the same, to the accuracy with which the standard deviations are usually quoted. Note that the square of the standard deviation is termed the *variance*.

Consider calculations of the standard deviations for the two sets of data used in Table 4.1. Table 4.2 shows the results of squaring the deviations.

Table 4.2 *Calculation of standard deviations*

1st set of readings		
Reading (s)	Deviation (s)	(Deviation)² s²
20.1	0.0	0.00
20.0	−0.1	0.01
20.2	+0.1	0.01
20.1	0.0	0.00
20.1	0.0	0.00

Sum of (deviation)² = 0.02, hence standard deviation = $\sqrt{\frac{0.02}{5-1}} = 0.07$,

2nd set of readings

Reading (s)	Deviation (s)	(Deviation)² s²
19.5	−0.6	0.36
20.5	+0.4	0.16
19.7	−0.4	0.16
20.6	+0.5	0.25
20.2	+0.1	0.01

Sum of (deviation)² = 0.94, hence standard deviation = $\sqrt{\frac{0.94}{5-1}} = 0.48$.

The second set of readings has a much greater standard deviation than the first set, indicating the greater spread of those results.

Note that many pocket calculators have a built-in facility for determining standard deviations without the need to carry out the individual steps outlined above. With my Casio calculator there are two standard deviation functions, labelled σ_n and σ_{n-1}, the first being, as indicated in the earlier discussion, for use with large numbers of measurements and the second with small numbers. The sequence of operations for calculating the standard deviations for the second set of measurements is:

1 Set the calculator to the statistics mode by, with my calculator, pressing SHIFT and MODE. The display then shows SD.

2 Key in the first reading, i.e. 19.5. Then press the M+ key.

3 Key in the second reading, i.e. 20.5. Then press the M+ key.

4 Repeat these steps for each reading.

5 Finally press the σ_{n-1} key. The standard deviation is then displayed. In this case 0.48.

It is important to realise that the standard deviation is a measure of the spread of the results. We can reasonably expect that about 68% of the readings will lie within plus or minus one standard deviation of the mean, 95% within plus or minus two standard deviations and 99.7% within plus or minus three standard deviations (see Chapter 5). Thus the standard deviation lets us know how far from the mean we can expect any one reading to be. Increasing the number of readings will have no noticeable effect on the standard deviation since the numerator and denominator of the equation for the standard deviation grow more or less proportionally. The spread of the overall set of results is independent of the sample size.

4.4.2 Error of a mean

If we take a set of readings and obtain a mean, how far from the true value might we expect the mean to be? Essentially what we do is consider the mean value of our set of results to be one of the many mean values which can be obtained from the very large number of results and calculate its standard deviation from that true value mean. To avoid confusion with the standard deviation of a single result from a mean of a set of results, we use the term *standard error* for the standard deviation of the mean from the true value. Thus, the extent to which we might expect a mean of a set of readings to be from the true value is given by the *standard error of the mean*, this being given by:

$$\text{standard error} = \frac{\text{standard deviation of the set of resul}}{\sqrt{n}}$$

Note that this equation is just stating that if we take n sets of readings, each with the same number of readings and the same standard deviation but different means, then (standard error)^2 is the average value of the square of the standard deviation of any one set.

We can reasonably expect that there is a 68% chance that a particular mean value lies within plus or minus one standard error of true value, a 95% chance within plus or minus two standard errors and a 99.7% chance within plus or minus three standard errors.

The greater the number of measurements made, the smaller will be the standard error. Because the factor is $\sqrt{n}$, increasing the number of readings taken by 100 only reduces the error by a tenth. Note that such a reduction in error is only a reduction in the random error, there may still be an unaffected systematic error.

Thus, for the data in Table 4.2, the first set of measurements had a standard deviation of 0.07 s and so a standard error of $0.07/\sqrt{5}$ = 0.03 s. Thus, with the mean value of 20.1 s, the result is 20.1 ± 0.03 s. The chance of the true value being within ±0.03 s of 20.1 is 68%. For the second set of measurements, the standard deviation was 0.48 s and so there is a standard error of $0.48/\sqrt{5}$ = 0.21 s. The result is then quoted as 20.1 ± 0.21 s. The chance of the true value being within ±0.21 s of the true value is 68%.

Example

Measurements of the electrical resistance of a resistor gave the following results: 53, 48, 45, 49, 46, 48, 51, 57, 55, 55, 47, 49 Ω. Determine the standard error.

The mean is the sum of the above values divided by the number of values involved, i.e.

$$\text{mean} = \frac{53+48+45+49+46+48+51+57+55+55+47+49}{12}$$

and so the mean is 603/12 = 50.25 Ω.

Table 4.3 illustrates the derivation of the standard deviation. The sum of (deviation)² = 168.25 and hence the standard deviation is √(168.25/11) = 3.91 Ω. The standard error is thus 3.91/√12 = 1.13 Ω and we can write our estimate of the resistance as 50.25 ± 1.13 Ω.

Table 4.3 *Derivation of the standard deviation*

Resistance (Ω)	Deviation (Ω)	(Deviation)² Ω²
53	+2.75	7.5625
48	−2.25	5.0625
45	−5.25	27.5625
49	−1.25	1.5625
46	−4.25	18.0625
48	−2.25	5.0625
51	+0.75	0.5625
57	+6.75	45.5625
55	+4.75	22.5625
55	+4.75	22.5625
47	−2.25	5.0625
49	−1.25	1.5625

4.5 Combining errors

An experiment might require several quantities to be measured and then the values inserted into an equation. For example, in a determination of the density ρ of a solid, measurements might be made of the mass m of the body and its volume V. Then the values are inserted into the equation:

$$\rho = \frac{m}{V}$$

in order to obtain the value for the density. The mass and volume measurements will each have errors associated with them. How then do we determine the consequential error in the density? This type of problem is very common. The following illustrates how we can determine the error in such situations.

4.5.1 The worst possible error

Consider the calculation of the quantity Z from two measured quantities A and B where $Z = A + B$. If the measured quantity A has an error $\pm \Delta A$ and the quantity B an error $\pm \Delta$ then the worst possible error we could have in Z is if the quantities are at the extremes of their error bands and the two errors are both positive or both negative. Then we have

$$Z + \Delta Z = (A + \Delta A) + (B + \Delta B)$$

$$Z - \Delta Z = (A - \Delta A) + (B - \Delta B)$$

Subtracting one equation from the other gives the worst possible error in Z as:

$$\Delta Z = \Delta A + \Delta B$$

When we add two measured quantities the worst possible error in the calculated quantity is the sum of the errors in the measured quantities.

If we have the calculated quantity Z as the difference between two measured quantities, i.e. $Z = A -$, then, in a similar way, we can show that the worst possible error is given by

$$Z + \Delta Z = (A + \Delta A) - (B + \Delta B)$$

$$Z - \Delta Z = (A - \Delta A) - (B - \Delta B)$$

and so, subtracting the two equations, gives the worst possible error as:

$$\Delta Z = \Delta A + \Delta B$$

When we subtract two measured quantities the worst possible error in the calculated quantity is the sum of the errors in the measured quantities.

If we have the calculated quantity Z as the product of two measured quantities A and B, I.e. $Z = A$, then we can calculate the worst error in Z as being when the quantities are both at the extremes of their error bands and the errors in A and B are both positive or both negative.

$$Z + \Delta Z = (A + \Delta A)(B + \Delta B) = AB + B\Delta A + A\Delta B + \Delta A \Delta$$

The errors in A and B are small in comparison with the values of A and B so we can neglect the quantity $\Delta A \Delta$ as being insignificant. Then

$$\Delta Z = B\Delta A + A\Delta B$$

Dividing through by Z gives:

$$\frac{\Delta Z}{Z} = \frac{B\Delta A + A\Delta B}{Z} = \frac{B\Delta A + A\Delta B}{AB} = \frac{\Delta A}{A} + \frac{\Delta B}{B}$$

Thus, when we have the product of measured quantities, the worst possible fractional error in the calculated quantity is the sum of the fractional errors in the measured quantities. If we multiply the above equation by 100 then we can state it as the: percentage error in Z is equal to the sum of the percentage errors in the measured quantities. If we have the square of a measured quantity, then all we have is the quantity multiplied by itself and so the error in the squared quantity is just twice that in the measured quantity. If the quantity is cubed then the area is three times that in the measured quantity.

If the calculated quantity is obtained by dividing one measured quantity by another, i.e. $Z = A/B$, then the worst possible error is given when we have the quantities at the extremes of their error bands and the error in A positive and the error in B negative, or vice versa. Then:

$$Z + \Delta Z = \frac{A + \Delta A}{B - \Delta B} = \frac{A\left(1 + \frac{\Delta A}{A}\right)}{B\left(1 - \frac{\Delta B}{B}\right)}$$

Using the Binomial series we can write this as:

$$Z + \Delta Z = \frac{A}{B}\left(1 + \frac{\Delta A}{A}\right)\left(1 + \frac{\Delta B}{B} - \ldots\right)$$

Neglecting products of ΔA and ΔB and writing A/B as Z, gives:

$$Z + \Delta Z = Z\left(1 + \frac{\Delta A}{A} + \frac{\Delta B}{B}\right)$$

Hence:

$$\frac{\Delta Z}{Z} = \frac{\Delta A}{A} + \frac{\Delta B}{B}$$

The worst possible fractional error in the calculated quantity is the sum of the fractional errors in the measured quantities or, if expressed in percentages, the percentage error in the calculated quantity is equal to the sum of the percentage errors in the measured quantities.

To sum up:

1 When measurements are added or subtracted, the resulting worst error is the sum of the errors.

2 When measurements are multiplied or divided, the resulting worst percentage errror is the sum of the percentage errors.

Example

The resistance R of a resistor is determined from measurements of the potential difference V across it and the current I through it. The resistance is given by V/I. The potential difference has been measured as 2.1 ± 0.2 V and the current measured as 0.25 ± 0.01 A, hence determine the worst possible error in the resistance.

The percentage error in the voltage reading is $(0.2/2.1) \times 100\% = 9.5\%$ and in the current reading is $(0.01/0.25) \times 100\% = 4.0\%$. Thus the percentage error in the resistance is $9.5 + 4.0 = 13.5\%$. Since we have $V/I = 8.4 \, \Omega$ and 13.5% of 8.4 is 1.1, then the resistance is $8.4 \pm 1.1 \, \Omega$.

Example

The cross-sectional area A of a wire is to be determined from a measurement of the diameter d, being given by $A = \pi d^2/4$. If the diameter is measured as 2.5 ± 0.1 mm, determine the area and its worst possible error.

The percentage error in d^2 is twice the percentage error in d. Since the percentage error in d is $\pm 4\%$ then the percentage error in d^2, and hence A since the others are pure numbers, is $\pm 8\%$. Since $\pi d^2/4 = 4.9$ mm² and 8% of this value is 0.4 mm², the result can be quoted as 4.9 ± 0.4 mm².

Example

The acceleration due to gravity g is determined from a measurement of the length L of a simple pendulum and the periodic time T, using:

$$T = 2\pi \sqrt{\frac{L}{g}}$$

If we have $L = 1.000 \pm 0.005$ m and $T = 2.0 \pm 0.1$ s, what is the value of the acceleration due to gravity and its worst possible error?

Squaring the equation and rearranging it gives:

$$g = \frac{4\pi^2 L}{T^2}$$

To give the percentage error in g, we need to add the percentage errors in L and T^2. The percentage error in L is 0.5% and that in T is 5.0%. Thus the total percentage error in g is $0.5 + 2 \times 5.0 = 10.5\%$. Putting the values of L and T in the equation gives a value of 9.87 m/s² and since 10.5% of this is 1.04 m/s² we can quote the result of the measurement, to two significant figures, as 9.9 ± 1.0 m/s².

Example

The coefficient of viscosity η of a liquid can be determined by measuring the rate of flow Q through a tube of radius r and length L when there is a pressure difference p between the ends.

$$\eta = \frac{\pi p r^4}{8LQ}$$

Determine an equation giving the worst possible error in the viscosity in terms of the errors in p, r, L and Q.

We need to add the fractional errors in the quantities in order to give the fractional error in the viscosity. Thus:

$$\text{fractional error in } \eta = \frac{\Delta p}{p} + 4\frac{\Delta r}{r} + \frac{\Delta L}{L} + \frac{\Delta Q}{Q}$$

4.5.2 The standard error

In the above discusssion, the worst possible errors were determined. Thus, when we added two quantities, the errors were assumed to be both at the extremes of their error bands and both positive or both negative and so give the maximum possible error. In practice it is likely that this will give an overestimate of the error. For example, consider measurements taken of two lengths, say 10 measurements of each, to give mean values with standard errors of 25 ± 1 mm for one and 40 ± 2 mm for the other (Figure 4.11 illustrates this with just three measurements). We can combine any one of the measurements of the first length with any one of the measurements of the second length to obtain the sum of the lengths. With 10 measurements of each, this will give 100 values for the sum. The worst possible error in the sum of the two measurements is when we use just the extreme measurements of each length, i.e. 25 + 1 mm and 40 + 2 mm to give 65 + 3 mm, and 25 − 1 mm and 40 − 2 mm to give 65 − 3 mm and so 65 ± 3 mm. This is only the error we get in the worst situation, generally the sum of a pair of the measurements will have a smaller error. We can obtain a more realistic error by considering all the 100 possible sums of the two quantities.

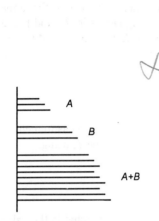

Figure 4.11 *The possible sums from 3 measurements of A and B*

Consider the error in a value of Z when we have $Z = A + B$ and there are errors $\pm \Delta A$ and $\pm \Delta B$ in the two quantities added. The error in a result is the difference between the calculated value Z and its true value $\bar{Z}$, i.e. $\Delta Z = Z - \bar{Z}$. We can write this as:

$$(\Delta Z)^2 = (Z - \bar{Z})^2$$

But $Z = A + B$ and $\bar{Z} = \bar{A} + \bar{B}$. Thus

$$(\Delta Z)^2 = [(A + B) - (\bar{A} + \bar{B})]$$

This can be rewritten as:

$$(\Delta Z)^2 = (A - \bar{A})^2 + (B - \bar{B})^2 + 2(A - \bar{A})(B - \bar{B})$$

The above equation gives the error for one combination of a measurement of A with one measurement of B and assumes that the errors are both the same sides of their true values, i.e. both errors are positive. However, the errors may be above or below their mean values and may be small or large. If we now consider all the possible combinations of A and B from each set of results, then we will have an equation of the above form for each set of errors. If we consider all the possible cases, add them together and then divide by the number of cases used we can obtain the standard error of Z. We would expect to find the $2(A - \bar{A})(B - \bar{B})$ term having the average value of 0. Thus on average we would expect:

$$(\Delta Z)^2 = (A - \bar{A})^2 + (B - \bar{B})$$

and so:

$$(\Delta Z)^2 = (\Delta A)^2 + (\Delta B)^2$$

The same expression is obtained when $Z = A - B$.

The following is another way of arriving at the same result. The worst error in Z, when both errors are positive and each quantity is at the extreme end of its error band, is given by (see previous section):

$$\Delta Z = \Delta A + \Delta B$$

Consider that each quantity, A and B, has been arrived at from a set of measurements of each and the mean and standard error obtained. If we consider all the possible sums of the quantities A and B, one or both of the errors could be negative or not at the extremes of the error band. Writing the equation in the form:

$$(\Delta Z)^2 = (\Delta A + \Delta B)^2$$

$$= (\Delta A)^2 + (\Delta B)^2 + 2\Delta A \Delta B$$

then adding together all the possible values the errors could take and dividing by the number considered, we would expect the $2\Delta A\Delta B$ term to have the value of zero since there will be as many situations with it having a negative value as having a positive value. Thus

$$(\Delta Z)^2 = (\Delta A)^2 + (\Delta B)^2$$

The same expression is obtained when $Z = A - B$.

Now consider the standard error when we have $Z = AB$ with errors $\pm \Delta A$ and $\pm \Delta B$. The worst error in Z is when the errors in A and B are both positive, or both negative. Then (see previous section):

$$\frac{\Delta Z}{Z} = \frac{\Delta A}{A} + \frac{\Delta B}{B}$$

Squaring both sides of the equation gives:

$$\left(\frac{\Delta Z}{Z}\right)^2 = \left(\frac{\Delta A}{A} + \frac{\Delta B}{B}\right)^2$$

$$= \left(\frac{\Delta A}{A}\right)^2 + \left(\frac{\Delta B}{B}\right)^2 + 2\left(\frac{\Delta A}{A}\right)\left(\frac{\Delta B}{B}\right)$$

But one or both of the errors could be negative or not at the extremes of the error band. Adding all the possible values the errors could take and dividing by the number considered, we would expect the $?(\Delta A/A)(\Delta B/B)$

term to have the value of zero since there will be as many situations with it having a negative value as having a positive value. Thus

$$\left(\frac{\Delta Z}{Z}\right)^2 = \left(\frac{\Delta A}{A}\right)^2 + \left(\frac{\Delta B}{B}\right)^2$$

The above equation involves the fractional errors, it could equally well be in terms of the percentage errors. The same equation is obtained when we have $Z = A/B$.

To sum up:

1. When measurements are added or subtracted, the resulting square of the standard error is the sum of the squares of the errors.

2. When measurements are multiplied or divided, the resulting square of the standard percentage errror is the sum of the squares of the percentage errors.

For an equation involving a power, e.g. $Z = AB^n$, then the error is given by:

$$\left(\frac{\Delta Z}{Z}\right)^2 = \left(\frac{\Delta A}{A}\right)^2 + n\left(\frac{\Delta B}{B}\right)^2$$

To illustrate the above, consider the examples used in the previous section for the determination of the worst errors.

Example

The resistance R of a resistor is determined from measurements of the potential difference V across it and the current I through it. The resistance is given by V/I. The potential difference has been measured as 2.1 ± 0.2 V and the current measured as 0.25 ± 0.01 A. Determine the resistance and the standard error.

The percentage error in the voltage reading is $(0.2/2.1) \times 100\% = 9.5\%$ and in the current reading is $(0.01/0.25) \times 100\% = 4.0\%$. Thus the percentage error in the resistance is given by:

$$(\% \text{ error in } R)^2 = 9.5^2 + 4.0^2$$

Hence the percentage error is 10.3%. Since $V/I = 8.4$ Ω and 10.3% of 8.4 is 0.87, then the resistance is 8.4 ± 0.9 Ω.

Example

The cross-sectional area A of a wire is determined from a measurement of the diameter d. We have $A = \pi d^2/4$. The diameter is measured as 2.5 ± 0.1 mm. Determine the area and its standard error.

Since the percentage error in d is ±4% then the percentage error in d^2, i.e. d multiplied by d, is given by:

$$(\% \text{ error in } d^2)^2 = 4^2 + 4^2$$

and so the percentage error is 5.7%. Since $\pi d^2/4 = 4.9$ mm² and 5.7% of this value is 0.3 mm², then the result can be quoted as 4.9 ± 0.3 mm².

Example

The acceleration due to gravity g can be determined from a measurement of the length L of a simple pendulum and the periodic time T. The relationship is:

$$g = \frac{4\pi^2 L}{T^2}$$

If we have $L = 1.000 \pm 0.005$ m and $T = 2.0 \pm 0.1$ s, determine the acceleration and the standard error.

The percentage error in L is 0.5% and that in T is 5.0%. Thus the total percentage error in g is given by:

$$(\% \text{ error in } g)^2 = 0.5^2 + 2(5.0)^2$$

Hence the error is 7.1%. Putting the values of L and T in the equation gives a value of 9.87 m/s² and since 7.1% of this is 0.70 m/s² we can quote the result to two significant figures as 9.9 ± 0.7 m/s².

Example

The coefficient of viscosity η of a liquid can be determined by measuring the rate of flow Q through a tube of radius r and length L when there is a pressure difference p between the ends, using the equation:

$$\eta = \frac{\pi p r^4}{8LQ}$$

Determine an equation giving the fractional standard error in the viscosity in terms of the fractional errors in the other quantities.

Using the equations developed above for products and quotients of quantities, the fractional standard error in the coefficient of viscosity is related to the other errors by:

$$\left(\frac{\Delta \eta}{\eta}\right)^2 = \left(\frac{\Delta p}{p}\right)^2 + 4\left(\frac{\Delta r}{r}\right)^2 + \left(\frac{\Delta L}{L}\right)^2 + \left(\frac{\Delta Q}{Q}\right)^2$$

4.5.3 Partial differentiation and the worst error

In the above analysis we have considered specific cases where we are required to sum two measurements, or subtract, or multiply, or divide. In general, if we want to find out how a small change in one quantity affects another, we differentiate. For example, if we have $Z = A^2$ then:

$$\frac{dZ}{dA} = 2A$$

If we consider finite elements, then we can write:

$$\Delta Z = 2A \Delta A$$

Dividing both sides of the equation by Z gives, since $Z = A^2$:

$$\frac{\Delta Z}{Z} = 2 \frac{\Delta A}{A}$$

Thus the fractional error in Z is twice the fractional error in A.

Now consider the situation where Z depends on more than one variable, e.g. $Z = A + B$. As before we can differentiate. However, when we have more than one variable, ordinary differentiation is replaced by *partial differentiation*. This involves differentiating assuming that only one variable is allowed to vary at a time, the others being temporarily constant. Thus, taking A to vary and B to be constant:

$$\frac{\partial Z}{\partial A} = 1$$

Note that partial derivatives are always written using a curly ∂ instead of the straight d used with ordinary derivatives. Taking B to vary and A to be constant:

$$\frac{\partial Z}{\partial B} = 1$$

As before, we can consider finite elements of the quantities. Thus when A varies and B is constant:

$$\Delta Z = \Delta A$$

and when B varies and A is constant:

$$\Delta Z = \Delta B$$

When we have a situation where both A and B can vary, then:

$$\Delta Z = \Delta A + \Delta B$$

This is just the equation we arrived at earlier.

In general, the equation we are using is:

$$\Delta Z = \frac{\partial Z}{\partial A}\Delta A + \frac{\partial Z}{\partial B}\Delta B$$

This equation can be expanded to as many terms as we require.

Example

Consider $Z = A\,e^B$ and determine an equation for the worst error in Z when there are errors in A and B.

Taking B to be constant:

$$\frac{\partial Z}{\partial A} = e^B$$

Taking A to be constant:

$$\frac{\partial Z}{\partial B} = A\,e^B$$

Thus the error in Z is given by:

$$\Delta Z = e^B\,\Delta A + A\,e^B\,\Delta B$$

Example

Consider a measurement of the refractive index n in which the angle of incidence i of a beam of light and its angle of refraction r are measured. We then have:

$$n = \frac{\sin i}{\sin r}$$

Determine the worst error in n when there are errors in i and r.

Taking r to be constant:

$$\frac{\partial n}{\partial i} = \frac{\cos i}{\sin r}$$

Taking i to be constant:

$$\frac{\partial n}{\partial r} = \frac{-\sin i \cos r}{\sin^2 r}$$

When we are determing the error with both quantities varying, we take just the magnitude of the partial differential, ignoring any minus signs that may occur. This avoids cancellation of one error by another. Thus

$$\Delta Z = \frac{\cos i}{\cos r}\Delta i + \frac{\sin i \cos r}{\sin^2 r}\Delta r$$

Example

The focal length f of a lens can be determined from measurements of the object distance u and the image distance v, with:

$$\frac{1}{f} = \frac{1}{u} + \frac{1}{v}$$

Determine an equation for the worst possible error in the focal length when there are errors in the object distance and the image distance.

The equation can be rearranged to give:

$$f = \frac{uv}{u+v}$$

Thus:

$$\frac{\partial f}{\partial u} = \frac{(u+v)v - uv}{(u+v)^2}$$

and

$$\frac{\partial f}{\partial v} = \frac{(u+v)u - uv}{(u+v)^2}$$

Hence:

$$\Delta f = \left(\frac{(u+v)v - uv}{(u+v)^2}\right)\Delta u + \left(\frac{(u+v)u - uv}{(u+v)^2}\right)\Delta v$$

4.5.4 Partial differentiation and the standard error

The above analysis gives the worst possible errors. To obtain the standard errors we have to square and average the square over a set of measurements. Taking the square of the error gives:

$$(\Delta Z)^2 = \left(\frac{\partial Z}{\partial A}\Delta A + \frac{\partial Z}{\partial B}\Delta B\right)^2$$

and hence:

$$(\Delta Z)^2 = \left(\frac{\partial Z}{\partial A}\right)^2 (\Delta A)^2 + \left(\frac{\partial Z}{\partial B}\right)^2 (\Delta B)^2 + 2\frac{\partial Z}{\partial A}\Delta A \frac{\partial Z}{\partial B}\Delta B$$

We can write such an equation for each measurement. Now consider all the measurements and the average error. The sum of all the last terms will be zero since ΔA and ΔB will have both positive and negative values. Thus, the standard error is given by:

$$(\Delta Z)^2 = \left(\frac{\partial Z}{\partial A}\right)^2 (\Delta A)^2 + \left(\frac{\partial Z}{\partial B}\right)^2 (\Delta B)$$

Example

Determine an equation for the standard error for the total resistance R of two resistances R_1 and R_2 in parallel due to errors in each.

$$\frac{1}{R} = \frac{1}{R_1} + \frac{1}{R_2}$$

We can write this as:

$$R = \frac{R_1 R_2}{R_1 + R_2}$$

This gives:

$$\frac{\partial R}{\partial R_1} = \frac{R_2(R_1 + R_2) - R_1 R_2}{(R_1 + R_2)^2}$$

$$\frac{\partial R}{\partial R_2} = \frac{R_1(R_1 + R_2) - R_1 R_2}{(R_1 + R_2)^2}$$

Hence:

$$(\Delta R)^2 = \left[\frac{R_2(R_1 + R_2) - R_1 R_2}{(R_1 + R_2)^2}\right]^2 (\Delta R_1)$$

$$+ \left[\frac{R_1(R_1 + R_2) - R_1 R_2}{(R_1 + R_2)^2}\right]^2 (\Delta R_2)$$

Problems

1. A thermometer has graduations at intervals of 0.5°C. What is the worst possible reading error?

2. An instrument has a scale with graduations at intervals of 0.1 units. What is the worst possible reading error?

3. Determine the means and the standard deviations for the following sets of results:

(a) The times taken for 10 oscillations of a simple pendulum: 51, 49, 50, 49, 52, 50, 49, 53, 49, 52 s.

(b) The diameter of a wire when measured at a number of points using a micrometer screw gauge: 2.11, 2.05, 2.15, 2.12, 2.16, 2.14, 2.16, 2.17, 2.13, 2.15 mm.

(c) The volume of water passing through a tube per 100 s time interval when measured at a number of times: 52, 49, 54, 48, 49, 49, 53, 48, 50, 53 cm^3.

4 Repeated measurements of the forces necessary to break a tensile test specimen gave: 802, 799, 800, 798, 801 kN. Determine (a) the average force, and (b) the standard error of the mean.

5 Repeated measurements of the resistance of a resistor gave: 51.1, 51.3, 51.2, 51.3, 51.7, 51.0, 51.5, 51.3, 51.2, 51.4 Ω. Determine (a) the average resistance, and (b) the standard error of the mean.

6 Repeated measurements of the voltage necessary to cause the breakdown of a dielectric gave: 38.9, 39.3, 38.6, 38.8, 38.8, 39.0, 38.7, 39.4, 39.7, 38.4, 39.0, 39.1, 39.1, 39.2 kV. Determine (a) the average breakdown voltage, and (b) the standard error of the mean.

7 The total resistance of two resistors in series is the sum of their resistances. Determine the worst possible error in the total resistance if the resistors are 50 Ω with 10% accuracy and 100 Ω with 5% accuracy.

8 When two resistors are connected in parallel, the total resistance is given by $R = R_1R_2/(R_1 + R_2)$. What will be the worst possible error in the total resistance if the resistors are 50 Ω with a 10% accuracy and 100 Ω with a 5% accuracy?

9 The volume of a cube with sides of L is L^3. If the length is measured as 121 ± 2 mm, determine the worst possible error in the volume.

10 The density of a solid is its mass divided by its volume. If the mass is measured as 42.5 ± 0.5 g and the volume as 54 ± 1 cm^3, what will be the worst possible error in the calculated density?

11 Two objects are weighed, giving 100 ± 0.5 g and 50 ± 0.3 g. What will be the average error of the sum of the two weights?

12 The distance s travelled by a car when travelling with a constant speed v is estimated from a reading of the speedometer and a measurement of time t travelled at that speed. If $s = vt$, what will be the distance travelled and the average error if the speed is 60 ± 2 km/h and the time is 1 ± 0.01 h?

13 The volume of a rectangular solid is determined from measurements of its length, height and breadth. If they are 100 ± 1 mm, 50 ± 0.5 mm and 40 ± 0.5 mm, what will be the volume and the average error?

14 The stress σ acting on a rectangular cross-section strip of material is determined from measurements of the force F and the cross-sectional area A, where $\sigma = F/A$. What will be the stress and the average error if the force is 20.0 ± 0.5 kN and the area is determined from measurements of the width and breadth of the strip as 5.0 ± 0.5 mm and 10.0 ± 0.5 mm?

15 The area S of a triangle is determined using the equation $S = \frac{1}{2} bc \sin A$. Determine an equation for the worst possible error in the area from measurements of the sides b and c and the angle A.

16 The number of active nuclei N in a sample of radioactive material after a time t is given by $N = N_0 e^{-\lambda t}$, where N_0 is the number at time $t = 0$. Determine an equation for the worst possible error in N due to errors in N_0 and t.

17 The viscosity η of a liquid can be determined from measurements of the terminal velocity v of a sphere falling through the liquid, with:

$$\eta = \frac{2r^2 g(\rho_s - \rho_l)}{9v}$$

where r is the radius of the sphere, g the acceleration due to gravity, ρ_s the density of the sphere and ρ_l the density of the liquid. Determine an equation for the worst possible error in the viscosity due to errors in r and v, g and the densities being assumed to have insignificant errors.

5 Statistics and data

All measurements are affected by random uncertainties and thus repeated measurements will give readings which fluctuate in a random manner from each other. This chapter is a consideration of the statistical approach to such variability of data, dealing with the background to the standard deviation and standard error (introduced in Chapter 4) and the confidence with which data can be considered.

5.1 Distributions

Consider some experiment in which repeated measurements are made of some quantity, e.g. the time taken for 100 oscillations of a simple pendulum. Suppose we take 20 readings and have the following results:

20.1, 20.3, 20.8, 20.5, 21.0, 20.8, 20.3, 20.4, 20.7, 20.6,
20.5, 20.7, 20.5, 20.1, 20.6, 20.4, 20.7, 20.5, 20.6, 20.3

We can display this data pictorially by means of a *histogram*. Thus, if we divide the data range into a number of convenient, equally sized, segments of, in this case, 0.2 we have:

values between 20.0 and 20.2 (>20.0 and ≤20.2) come up twice
values between 20.2 and 20.4 (>20.2 and ≤20.4) come up five times
values between 20.4 and 20.6 (>20.4 and ≤20.6) come up seven times
values between 20.6 and 20.8 (>20.6 and ≤20.8) come up five times
values between 20.8 and 21.0 (>20.8 and ≤21.0) come up once

The term *frequency* is used for the number of times a measurement occurs within a segment and the histogram represents the *frequency distribution*, showing how the data values group together. Figure 5.1 shows the histogram for the above data. The horizontal axis gives the measurement values for the mid points of each segment.

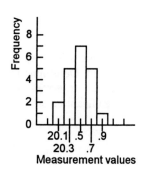

Figure 5.1 *Histogram of the data*

Using the frequency, it is difficult to compare a histogram obtained with the 20 values with one obtained with, say, 100 values. For this reason, the *relative frequency* tends to be used. The relative frequency is the fraction of the total number of readings in a segment. Thus, for the data given above we have:

values between 20.0 and 20.2, relative frequency 2/20 = 0.1
values between 20.2 and 20.4, relative frequency 5/20 = 0.25
values between 20.4 and 20.6, relative frequency 7/20 = 0.35
values between 20.6 and 20.8, relative frequency 5/20 = 0.25
values between 20.8 and 21.0, relative frequency 1/20 = 0.05

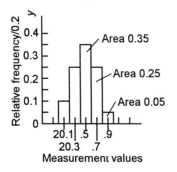

Figure 5.2 *Histogram with relative frequency*

The relative frequency always has a value less than 1 and the sum of all the relative frequencies is 1, since this is the relative frequency for all the readings. The area of a rectangular segment in the histogram is made equal to relative frequency for that segment and hence the total area of all the strips is 1. Thus we can write for the relative frequency of a segment of width Δx and ordinate y:

$$\text{relative frequency} = y \, \Delta x$$

Figure 5.2 shows the above results with the vertical axis as the relative frequency per unit segment width, in this case the relative frequency per 0.20. The first segment has an ordinate y of $(0.1/0.2)$ and thus an area of $(0.1/0.2) \times 0.20 = 0.1$. A distribution plotted using relative frequencies per unit segment is said to be *normalised*.

The higher the relative frequency of a segment, the greater the probability that if we take a single measurement at random from the entire set that it will lie in that segment. The *probability* of an event occurring is the frequency with which it occurs as a fraction of the total number of possible outcomes.

$$\text{probability} = \frac{\text{number of ways an event can occur}}{\text{total number of ways possible}}$$

A coin can land either heads or tails uppermost. There are thus two possible outcomes of tossing a coin. The probability of a single toss of a coin landing heads uppermost, i.e. just one way out of the possible two, is 1 in 2 or 0.5. The relative frequency may thus be considered to be the probability. Thus, for the segment 20.6 to 20.8 there is a relative frequency of 0.25. This means the probability of a single measurement having a value between 20.6 and 20.8 is 0.25, i.e. 1 in 4. Of the total 20 values, 1 in 4, i.e. 5, lie between 20.6 and 20.8.

The histogram shown in Figure 5.2 has a jagged appearance. This is because it represents only a few values. If we had taken a very large number of readings then we could have divided the range into smaller segments and still had an appreciable number of values in each segment. The result of plotting the histogram would now be to give one with a much smoother appearance. As the number of readings used increases, not only does the histogram become smoother, but it settles down into a constant shape. For example, the normalised histograms for 10 and 20 readings will be jagged and could vary quite significantly in their general shape. However, the normalised histograms for 500 and 1000 readings are likely to be smooth with virtually no differences in their shape. Thus, with such large numbers, we might have a distribution of the form shown in Figure 5.3, a smooth curve having been drawn through the tops of the infinitely small segments of the histogram. The y axis is generally termed the *frequency function*, being the relative frequency per unit segment when the segments are infinitesimally small. This distribution represents the *limiting frequency distribution* we would obtain with an infinite number of values.

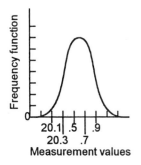

Figure 5.3 *Distribution with large numbers of readings*

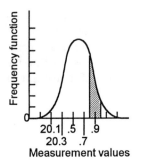

Figure 5.4 *Probability of a single measurement of 20.9*

If we consider a segment of a limiting frequency distribution then the relative frequency of that segment, i.e. its area, gives the *probability* that a single measurement taken at random from the distribution will lie in that segment. Consider the probability, with a very large number of readings, of obtaining a value between 20.8 and 21.0 with the distribution shown in Figure 5.4. If we take a segment 20.8 to 21.0 then the area of that segment is the relative frequency. Suppose the result in this case is 0.30. The probability of taking a single measurement and finding it in that interval is 0.30, i.e. on average 30 times in every 100 values taken.

Example

The following readings, in metres, were made for a measurement of the distance travelled by an object in 10 s. Plot the results as a normalised histogram with segments of width 0.01 m.

13.478, 13.509, 13.502, 13.457, 13.492, 13.512, 13.475, 13.504, 13.473, 13.482, 13.492, 13.500, 13.493, 13.501, 13.472, 13.477

With segments of width 0.01 m we have:

Segment 13.45 to 13.46, frequency 1 and so relative frequency 1/16
Segment 13.46 to 13.47, frequency 0 and so relative frequency 0
Segment 13.47 to 13.48, frequency 5 and so relative frequency 5/16
Segment 13.48 to 13.49, frequency 1 and so relative frequency 1/16
Segment 13.49 to 13.50, frequency 4 and so relative frequency 4/16
Segment 13.50 to 13.51, frequency 4 and so relative frequency 4/16
Segment 13.51 to 13.52, frequency 1 and so relative frequency 1/16

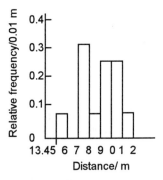

Figure 5.5 *Example*

Figure 5.5 shows the resulting histogram.

5.1.1 Mean

The mean value $\bar{x}$ of a set of readings can be obtained in a number of ways, depending on the form with which the data is presented:

1. For a list of discrete readings, sum all the readings and divide by the number N of readings, i.e.:

$$\bar{x} = \frac{x_1 + x_2 + x_3 + \ldots + x_j}{N} = \frac{\sum x_j}{N}$$

2. For a distribution of discrete readings (as, for example, in Figure 5.6), if we have n_1 readings with value x_1, n_2 readings with value x_2, n_3 readings with value x_3, etc., then the above equation for the mean becomes:

$$\bar{x} = \frac{n_1 x_1 + n_2 x_2 + n_3 x_3 + \ldots + n_j x_j}{N}$$

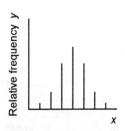

Figure 5.6 *Discrete distribution*

But n_1/N is the relative frequency of value x_1, n_2/N is the relative frequency of value x_2, etc. Thus, to obtain the mean, multiply each reading by its relative frequency y and sum over all the values.

$$\bar{x} = \sum_{j=1}^{n_j} y_j x_j$$

3   For readings presented as a histogram plotted with relative frequencies (Figure 5.7), the above relationship translates into: the mean is given by multiplying the mid-ordinate of each measurement value segment by the relative frequency for that segment and summing over all the possible values. Thus if we consider segment j in the histogram then it has a mid-ordinate measurement value of x_j and a relative frequency of $y_j \Delta x$. Thus the mean is given by:

$$\bar{x} = \sum_{j=1}^{n_j} y_j x_j \Delta x$$

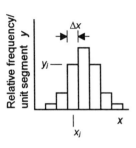

Figure 5.7 *Histogram*

4   For readings presented as a continuous distribution curve, we can consider that we have a histogram with very large numbers of very thin segments. Thus if y, a function of x, represents relative frequency values and x the measurement values, the rule given above for histograms translates into:

$$\bar{x} = \int_{-\infty}^{\infty} xy\, dx \text{ or } \int_{-\infty}^{\infty} xf(x)\, dx$$

For a symmetrical distribution, as in Figure 5.4, the mean will be the value with the greatest frequency.

With a very large number of readings, the mean value is taken as being the *true value* about which the random fluctuations occur.

Example

In the measurement of the speed of light in 1931 by A.A. Michelson, F.G. Peace and F. Pearson a large number of measurements were made and, in order to see how the results were distributed, the velocities were grouped into velocity ranges with a span of 5 km/s. Table 5.1 shows the results quoted in their paper *(Astrophysical Journal*, volume 82, 1935). Determine the mean velocity.

74 Experimental Methods

Table 5.1 *Speed of light data*

Velocity range km/s	Number of times	Velocity range km/s	Number of times
299 000 +		299 000 +	
726 to 730	4	776 to 780	515
731 to 735	6.5	781 to 785	270
736 to 740	3.0	786 to 790	236
741 to 745	55	791 to 795	90
746 to 750	29	796 to 800	62
751 to 755	86	801 to 805	33
756 to 760	184	806 to 810	30
761 to 765	304	811 to 815	32.5
766 to 770	353.5	816 to 820	0
771 to 775	580	821 to 825	12

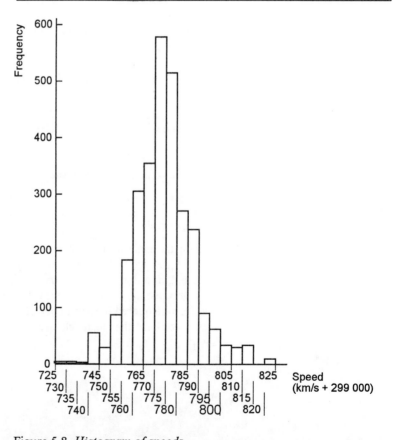

Figure 5.8 *Histogram of speeds*

Figure 5.8 shows the histogram of the above results. The mean for a histogram is obtained by multiplying the mid value of the measurement in each interval by the relative frequency for that interval and summing over all the possible values. Each interval has a width of 5 km/s and the total number of measurements listed in the table is 2885.5. Thus, for the interval 299 726 to 299 730 km/s the relative frequency is 4/2885.5 and so for that segment of the histogram we have 299 727.5 × 4/2885.5 = 415 km/s. We can repeat this calculation for each of the intervals and then sum all the results. The result is a mean of 299 774 km/s.

5.2 Standard deviation

Any single reading x in a distribution (Figure 5.9) will deviate from the mean of that distribution by some error e, i.e.:

$$e = x - \bar{x}$$

With a particular measurement we might obtain a series of readings which is widely scattered around the mean while another has readings closely grouped round the mean. The measurement with the large random fluctuations will give a distribution which is a broad peaked distribution curve while that with smaller random fluctuations will give a narrow peaked curve. Figure 5.10 shows the type of curves that might occur. The broad peaked distribution curve is said to be for a more imprecise set of readings than the narrow peaked curve. A measurement is said to be *precise* when it is determined from readings in which the random errors are small. This is irrespective of whether or not systematic errors are present. An accurate measurement is when it is determined from readings in which the random and systematic errors are small.

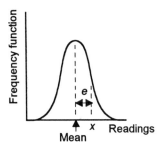

Figure 5.9 *Error e*

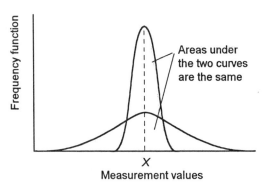

Figure 5.10 *Distributions with different precisions*

A measure of the precision can thus be obtained by some measure of the 'width' of the distribution curves. The average error cannot be used, since for every positive value of an error there will be a negative error and so the sum of all the errors will be zero. The measure used is the standard deviation. The *standard deviation* σ is the root-mean-square value of e for all the measurements in the distribution, i.e. σ^2 is the mean value of e^2. The quantity σ^2 is known as the *variance* of the distribution. Thus, for a number of discrete values, $x_1, x_2, x_3, \ldots$, etc., we can write for the mean value of the sum of the squares of their deviations from the mean of the set of results:

$$\text{sum of squares of deviation} = \frac{(x_1 - \bar{x})^2 + (x_2 - \bar{x})^2 + (x_3 - \bar{x})^2 + \ldots}{N}$$

Hence the mean of the square root of this sum of the squares of the deviations, i.e. the standard deviation, is:

$$\sigma = \sqrt{\frac{\left((x_1 - \bar{x})^2 + (x_2 - \bar{x})^2 + (x_3 - \bar{x})^2 + \ldots\right)}{N}}$$

However, we need to distinguish between the standard deviation of a sample s and the standard deviation σ of the entire population of readings that are possible and from which we have only considered a sample (many statistics textbooks adopt the convention of using Greek letters when referring to the entire population and Roman for samples). When we are dealing with a sample we need to write:

$$s = \sqrt{\frac{\left((x_1 - \bar{x}_s)^2 + (x_2 - \bar{x}_s)^2 + (x_3 - \bar{x}_s)^2 + \ldots\right)}{N-1}}$$

with $\bar{x}_s$ being the mean value of the sample rather than the true value $\bar{x}$. The reason for the $N-1$ rather than N is that in calculating the standard deviation we are using the sample mean rather than the true mean value that would have been used if we had used the entire population. The root-mean-square of the deviations of the readings in a sample around the sample mean is less than around any other figure. Hence, if the true mean of the entire population were known, the estimate of the standard deviation of the sample data about it would be greater than that about the sample mean. Therefore, by using the sample mean, an underestimate of the population standard deviation is given. This bias can be corrected by using one less than the number of observations in the sample in order to give the sample mean.

A form of this equation which can simplify calculations, by not requiring each deviation to be individually calculated, is obtained in the following way. We can write the above equation for the sample standard deviation s as:

$$s^2 = \frac{1}{N-1} \sum_{j=1}^{N} (x_j - \bar{x}_s)^2$$

where j has the values 1, 2, 3, ..., etc., up to N. Hence:

$$s^2 = \frac{1}{N-1} \sum_{j=1}^{N} \left(x_j^2 - 2\bar{x}_s x_j + \bar{x}_s^2 \right)$$

$$= \frac{1}{N-1} \left[\sum_{j=1}^{N} x_j^2 - \sum_{j=1}^{N} 2\bar{x}_s x_j + \sum_{j=1}^{N} \bar{x}_s^2 \right]$$

$$= \frac{1}{N-1} \left[\sum_{j=1}^{N} x_j^2 - 2\bar{x}_s \sum_{j=1}^{N} x_j + N\bar{x}_s^2 \right]$$

$$= \frac{1}{N-1} \left[\sum_{j=1}^{N} x_j^2 - 2\bar{x}_s N\bar{x}_s + N\bar{x}_s^2 \right]$$

$$= \frac{1}{N-1} \left[\sum_{j=1}^{N} x_j^2 - N\bar{x}_s^2 \right]$$

Hence:

$$s^2 = \frac{1}{N-1} \sum_{j=1}^{N} x_j^2 - \frac{N}{N-1} \bar{x}_s^2$$

If we were dealing with the entire population then the above equation would be:

$$\sigma^2 = \frac{1}{N} \sum_{j=1}^{N} x_j^2 - \bar{x}^2$$

i.e. the mean values of the squares minus the square of the mean.

When dealing with the entire population, if we have a number of discrete values, $x_1, x_2, x_3, \ldots$, etc., with frequencies n_1, n_2, n_3, etc., we can write:

$$\sigma^2 = \frac{(x_1 - \bar{x})^2 n_1 + (x_2 - \bar{x})^2 n_2 + (x_3 - \bar{x})^2 n_3 + \ldots}{N}$$

$$= \frac{1}{N} \sum_{j=1}^{N} n_j (x_j - \bar{x})^2$$

In the case of a sample, this equation becomes:

$$s^2 = \frac{1}{N-1} \sum_{j=1}^{N} n_j (x_j - \bar{x})^2$$

This equation can also be rearranged, as the equation for discrete values was, to the form:

$$s^2 = \frac{1}{N-1} \left[\sum_{j=1}^{N} n_j x_j^2 - N\bar{x}^2 \right]$$

When we have a histogram of the entire population with relative frequencies $y_1 \Delta x$, $y_2 \Delta x$, $y_3 \Delta x$, ... , etc., the number of times each value will occur is $Ny_1 \Delta x$, $Ny_2 \Delta x$, $Ny_3 \Delta x$, ... , etc., where N is the size of the population. Thus:

$$\sigma^2 = \frac{N}{N}[(x_1 - \bar{x})^2 y_1 \Delta x + (x_2 - \bar{x})^2 y_2 \Delta x + (x_3 - \bar{x})^2 y_3 \Delta x + ...]$$

$$= \sum_j^N (x_j - \bar{x})^2 y_j \Delta x$$

For a continuous distribution we can thus write:

$$\sigma^2 = \int_{-\infty}^{\infty} (x - \bar{x})^2 y \, dx \text{ or } \int_{-\infty}^{\infty} (x - \bar{x})^2 f(x) \, dx$$

where $y = f(x)$ is the frequency function for the distribution. We can write this equation in a more useful form for calculation. Thus expanding the equation:

$$\sigma^2 = \int_{-\infty}^{\infty} x^2 f(x) \, dx - 2\bar{x} \int_{-\infty}^{\infty} x f(x) \, dx + \bar{x}^2 \int_{-\infty}^{\infty} f(x) \, dx$$

Since the total area under the frequency function curve is 1, the third integral has the value 1 and so the third term in the equation is $\bar{x}^2$. The second integral is $\bar{x}$ and so the second term is $2\bar{x}^2$. The first integral is the mean value of x^2. Thus:

$$\sigma^2 = \overline{x^2} - \bar{x}^2$$

i.e. the mean value of x^2 minus the square of the mean value.

A distribution having a small value of σ will indicate a more precise measurement than one with a large value. The quantity denoted by σ is a measure of the error from the mean value of the sample we can expect for a single reading taken in that sample. In the case of a sample, we must not assume that the mean value of the sample is necessarily the true value that would occur if we had an infinitely large sample size. The standard deviation of a sample is a measure of the spread of the results in the sample, and this is not necessarily the same as the standard deviation of the population.

Example

Determine the mean value and the standard deviation of a sample of ten readings which gave the following results:

8, 6, 8, 4, 7, 5, 7, 6, 6, 4

The mean value is:

$$\bar{x}_s = \frac{8+6+8+4+7+5+7+6+6+4}{10} = 6.1$$

The standard deviation of the sample can be calculated by considering the deviations of each reading from the mean or using:

$$s^2 = \frac{1}{N-1}\sum_{j=1}^{N} x_j^2 - \frac{N}{N-1}\bar{x}_s^2$$

$$= \tfrac{1}{9}(8^2 + 6^2 + 8^2 + 4^2 + 7^2 + 5^2 + 7^2 + 6^2 + 6^2 + 4^2) - \tfrac{10}{9} \times 6.1^2$$

Hence the standard deviation is 1.4.

Example

The density of an acid in containers was sampled by measuring it for six containers. The results were:

1.843, 1.847, 1.842, 1.848, 1.845, 1.847 g/cm³

Determine the mean value of the density and the standard deviation of the sample.

The mean value of the sample is:

$$\bar{x}_s = \tfrac{1}{6}(1.843 + 1.847 + 1.842 + 1.848 + 1.845 + 1.847)$$

$$= 1.845(333) \text{ g/cm}^3$$

The standard deviation can be calculated from the individual deviations or by using:

$$s^2 = \frac{1}{N-1}\sum_{j=1}^{N} x_j^2 - \frac{N}{N-1}\bar{x}_s^2$$

$$= \tfrac{1}{5}(1.843^2 + 1.847^2 + 1.842^2 + 1.848^2 + 1.845^2 + 1.847^2) - \tfrac{6}{5}1.845333^2$$

Hence the standard deviation is 0.00027 g/cm³.

Rounding errors occur in the above calculations if 1.845 is used instead of 1.845333. If you used a calculator giving even more decimal places to directly determine the standard deviation, the value obtained would be 0.0024 g/cm³ (see sections 4.4 and 4.4.1).

Example

In an experiment involving the counting of the number of events, e.g. the detection of cosmic ray particles or perhaps alpha particles scattered at a particular angle) that occurred in equal size time intervals the following data was obtained:

0 events 13 times
1 event 12 times
2 events 9 times
3 events 5 times
4 events once

Determine the mean number of events occurring in the time interval and the standard deviation.

The total number of measurements made is $13 + 12 + 9 + 5 + 1 = 40$ and so the mean value is:

$$\bar{x} = \frac{0 \times 13 + 1 \times 12 + 2 \times 9 + 3 \times 5 + 4 \times 1}{40} = 1.25$$

Using the equation:

$$s^2 = \frac{1}{N-1}\left[\sum_{j=1}^{N} n_j x_j^2 - N\bar{x}^2\right]$$

then the standard deviation is given by:

$$s^2 = \tfrac{1}{39}[(13 \times 0^2 + 12 \times 1^2 + 9 \times 2^2 + 5 \times 3^2 + 1 \times 4^2) - 40 \times 1.25^2]$$

Hence the standard deviation is 1.1.

Example

A continuous frequency distribution is of the form shown in Figure 5.11. Determine its mean and standard deviation.

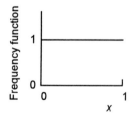

Figure 5.11 *Example*

For the distribution, $f(x) = 1$. The mean value for a distribution curve is given by:

$$\bar{x} = \int_{-\infty}^{\infty} xf(x)\,dx = \int_0^1 x\,dx = \left[\frac{x^2}{2}\right]_0^1 = \frac{1}{2}$$

The standard deviation is given by the equation:

$$\sigma^2 = \int_{-\infty}^{\infty}(x-\bar{x})^2 f(x)\,dx = \int_0^1 \left(x - \tfrac{1}{2}\right)^2 dx$$

$$= \int_0^1 \left(x^2 - x + \tfrac{1}{4}\right) dx = \left[\frac{x^3}{3} - \frac{x^2}{2} + \frac{x}{4}\right]_0^1$$

$$= \tfrac{1}{3} - \tfrac{1}{2} + \tfrac{1}{4} = \tfrac{1}{12}$$

Hence the standard deviation is 0.29.
Alternatively, we could have used the equation:

$$\sigma^2 = \overline{x^2} - \bar{x}^2$$

Since:

$$\overline{x^2} = \int_{-\infty}^{\infty} x^2 f(x)\,dx = \int_0^1 x^2\,dx = \left[\frac{x^3}{3}\right]_0^1 = \frac{1}{3}$$

then

$$\sigma^2 = \tfrac{1}{3} - \left(\tfrac{1}{2}\right)^2 = \tfrac{1}{3} - \tfrac{1}{4} = \tfrac{1}{12}$$

Hence, as before, the standard deviation is 0.29.

5.3 Standard error of the mean

When a set of readings is taken we can determine its mean, but what is generally required is an estimate of the error of that mean from the true value, i.e. the mean of an infinitely large number of readings. We can consider any set of readings as being just a sample taken from the very large set.

Consider a container holding six counters each having one of the numbers 3, 4, 5, 6, 7, 8. Now suppose we draw, at random, a sample of two of these counters. There are 15 different combinations of two counters we can draw. The following are the possible samples and their means:

3, 4	3, 5	3, 6	3, 7	3, 8	4, 5	4, 6	4, 7	4, 8	5, 6	5, 7	5, 8	6, 7	6, 8	7, 8
3.5	4.0	4.5	5.0	5.5	4.5	5.0	5.5	6.0	5.5	6.0	6.5	6.5	7.0	7.5

The mean values of the samples vary from 3.7 to 7.5. The mean of these means is 5.5 and the standard deviation of the means from 5.5 is 1.12. Now suppose that, instead of samples of two, we take samples of four. There are 15 different combinations of four counters we can draw. The following are the possible samples and their means:

3, 4, 5, 6	3, 4, 5, 7	3, 4, 5, 8	3, 4, 6, 7	3, 4, 6, 8	3, 5, 6, 7	3, 5, 6, 8
4.5	4.75	5.0	5.0	5.25	5.25	5.5

3, 5, 7, 8	3, 6, 7, 8	3, 4, 7, 8	4, 5, 6, 7	4, 5, 6, 8	4, 5, 7, 8	4, 6, 7, 8
5.75	6.0	5.5	5.5	5.75	6.0	6.25

5, 6, 7, 8
6.5

The mean values of the samples vary from 4.5 to 6.5. The mean of these means is 5.5 and the standard deviation of the means from 5.5 is 0.56. The standard deviation is half that occurring with the samples of two taken from the population. Increasing the sample size by a factor of 2 has halved the standard deviation. With the sample size of four, the means are thus much more closely bunched around the mean than with a sample size of two. The larger sample size thus reduces the chance of any one sample being too far from the mean.

The following is an algebraic analysis of the effect of sample size on the standard deviation of a mean from the true mean. Consider one sample of readings with n values; $x_1, x_2, x_3, \ldots x_n$. The error e_j in the jth reading is:

$$e_j = x_j - X$$

where X is the true value of the quantity. This true value is not known, being the value of the mean when the sample is of infinite size and so the total population. The mean $\bar{x}$ of this sample is:

$$\bar{x} = \frac{1}{n} \sum x_j$$

This mean will have an error E from the true value X of:

$$E = \bar{x} - X$$

Hence we can write:

$$E = \left(\frac{1}{n} \sum x_j\right) - X$$

This can be rewritten as:

$$E = \frac{1}{n} \sum (x_j - X)$$

and hence:

$$E = \frac{1}{n} \sum e_j$$

Thus what we have is:

$$E = \frac{1}{n}\left(e_1 + e_2 + e_3 + \ldots e_j\right)$$

Thus:

$$E^2 = \frac{1}{n^2}(e_1^2 + e_2^2 + e_3^2 + \ldots + \text{products such as } e_1 e_2, \text{ etc.})$$

E is the error from the mean for a single sample of readings. Now, consider a large number of such samples with each set having the same number n of readings. We can write such an equation as above for each sample. If we add together the equations for all the samples and divide by the number of samples considered, we obtain an average value over all the samples of E^2. Thus E is the standard deviation of the means and is known as the *standard error of the means* e_m (more usually the symbol σ). Adding together all the error product terms will give a total value of zero, since as many of the error values will be negative as well as positive. The average of all the Σe_j^2 terms is $n e_s^2$, where e_s is, what can be termed, the *standard error of the sample*. Thus:

$$e_m = \frac{e_s}{\sqrt{n}}$$

But how we can we obtain a measure of the standard error of the sample? The standard error is measured from the true value X, which is not known. What we can measure is the standard deviation of the sample from its mean value. The best estimate of the standard error for a sample turns out to be the standard deviation s of a sample when we define it as:

$$s^2 = \frac{1}{n-1} \Sigma (x_j - \bar{x})^2$$

i.e. with a denominator of $N - 1$, rather than just N. Thus the best estimate of the standard error of the mean can be written as:

$$\text{standard error of the mean } \sigma = \frac{s}{\sqrt{n}}$$

Example

Measurements are to be made of the percentage of an element in a chemical by making measurements on a number of samples. The standard deviation of any one sample is found to be 2%. How many measurements must be made to give a standard error of 0.5% in the estimated percentage of the element.

If n measurements are made, then the standard error of the sample mean is given by:

$$\text{standard error of the mean} = \frac{s}{\sqrt{n}}$$

Thus:

$$n = \frac{2^2}{0.5^2} = 16$$

16 measurements should be made.

84 *Experimental Methods*

Example

Measurements were made by two students of the tensile strength of a polymer, two different sets of samples being used. The results obtained by the students were:

Student A
29.0, 28.7, 28.9, 29.3, 28.8, 29.4, 29.1, 29.2, 29.7, 28.4, 29.0, 29.1, 28.6, 28.8 MPa

Student B
29.2, 29.8, 29.8, 29.4, 29.6, 28.9, 29.4, 29.5, 28.7, 29.6, 30.2, 29.2, 29.9, 29.9 MPa

Do the results indicate that the two sets of samples are from the same batch?

For the results from student A, using a calculator, the mean was found to be 29.0 MPa and standard deviation to be 0.339 MPa. The standard error of the mean, for the 14 samples, is thus $0.339/\sqrt{14} = 0.091$ MPa. For the results from student B the mean was found to be 29.5 MPa and the standard deviation to be 0.396 MPa. The standard error of the mean, for the 14 samples, is thus $0.396/\sqrt{14} = 0.106$ MPa. The difference between the means is 0.5 MPa and this is more than can be accounted for by the standard errors of the two results (Figure 5.12). Thus it seems unlikely that the two sets of samples came from the same batch. This type of problem will be considered in more detail later in the chapter when we discuss the probability of such a difference in means occurring.

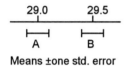

Figure 5.12 *Example*

5.4 Normal distribution

A particular form of distribution, known as the *normal distribution* or *Gaussian distribution*, works well as a model for measurements when there are random errors. For example, the distribution of the speed of light measurements given in the example with Figure 5.7 reasonably approximates to the form of the normal distribution. This form of distribution has a characteristic bell shape (Figure 5.13). It is symmetric about its mean value X, having its maximum value at that point. It tends rapidly to zero as x increases or decreases from the mean. It can be completely described in terms of its mean X and its standard deviation σ. The following equation describes how the values are distributed about the mean.

$$y = f(x) = \frac{1}{\sigma\sqrt{2\pi}} e^{-(x-X)^2/2\sigma^2}$$

Figure 5.13 *Typical forms of the normal distribution*

The fraction of the total number of measurements that lies between $-x$ and $+x$ from X is the fraction of the total area under the curve that lies

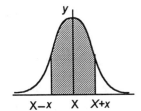

Figure 5.14 *The area between −x and +x of X*

between those ordinates (Figure 5.14). We can obtain areas under the curve by integration.

To save the labour of carrying out the integration, the results have been calculated and are available in tables. As the form of the graph depends on the value of the standard deviation, as illustrated in Figure 5.13, the area depends on the value of the standard deviation σ. In order not to have to give tables of the areas for different values of x for each value of σ, the distribution is considered in terms of the value of $(x - X)/\sigma$, this commonly being designated by the symbol z, and areas tabulated against this quantity. Table 5.2 shows examples of the type of data given in such tables:

Table 5.2 *Areas under normal curve*

$(x - X)/\sigma$	Area from X	$(x - X)/\sigma$	Area from X
0	0.0000	1.6	0.4452
0.2	0.0793	1.8	0.4641
0.4	0.1555	2.0	0.4772
0.6	0.2257	2.2	0.4861
0.8	0.2881	2.4	0.4918
1.0	0.3413	2.6	0.4953
1.2	0.3849	2.8	0.4974
1.4	0.4192	3.0	0.4987

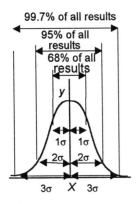

Figure 5.15 *Percentages of results in bands of 1, 2 and 3 standard deviations*

When $x - X = 1\sigma$, then $(x - X)/\sigma = 1.0$ and the area between the ordinate at the mean and the ordinate at 1σ, as a fraction of the total area, is 0.3413. The total area within $\pm 1\sigma$ of the mean is thus the fraction 0.6816. Expressed as a percentage, the area is 68.16%. This means that the chance of a reading being within $\pm 1\sigma$ of the mean is 68.16%, i.e. roughly two-thirds of the readings (Figure 5.15).

When $x - X = 2\sigma$, then $(x - X)/\sigma = 2.0$ and the area between the ordinate at the mean and the ordinate at 1σ, as a fraction of the total area, is 0.4772. The total area within $\pm 2\sigma$ of the mean is thus the fraction 0.9544. Expressed as a percentage, the area is 95.44%. This means that the chance of a reading being within $\pm 2\sigma$ of the mean is 95.44%.

When $x - X = 3\sigma$, then $(x - X)/\sigma = 3.0$ and the area between the ordinate at the mean and the ordinate at 3σ, as a fraction of the total area, is 0.4987. The total area within $\pm 1\sigma$ of the mean is thus the fraction 0.9974. Expressed as a percentage, the area is 99.74%. This means that the chance of a reading being within $\pm 3\sigma$ of the mean is 99.74%. Thus, virtually all the readings will lie within $\pm 3\sigma$ of the mean.

Example

Measurements of the diameter of a rod at a number of points give a mean of 2.500 cm and a standard deviation of 0.005 cm. What will be

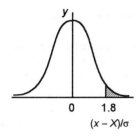

Figure 5.16 *Area >1.8*

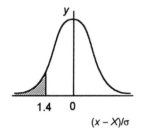

Figure 5.17 *Area <1.4*

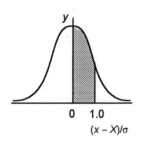

Figure 5.18 *Area between mean and 1.0*

the chance that a diameter reading will exceed 2.504 cm if five further measurements are made?

The means can be expected to follow a normal distribution. The standard error of the mean when 5 readings are taken is thus $0.005/\sqrt{5} = 0.0022$ cm. To use Table 5.2 we need to calculate $(x - X)/\sigma$. Thus we have $(2.504 - 2.500)/0.0022 = 1.8$. The table gives the area between the mean and this value as 0.4641. The area greater than the 1.8 value (Figure 5.16) is thus $0.500 - 0.4641 = 0.0469$ (the area between the mean and an extreme for the distribution is half the total area). Thus about 4.7% of the readings might be expected to have values greater than 2.504 cm. In the above analysis, it was assumed that the mean given was the true value or a good enough approximation.

Example

Measurements are made of the tensile strengths of samples taken from a batch of steel sheet. The mean value of the strength is 800 MPa and it is observed that 8% of the samples give values that are below an acceptable level of 760 MPa. What is the standard deviation of the distribution if it is assumed to be normal?

This means that an area from the mean of $0.50 - 0.08 = 0.42$. To the accuracy given in Table 5.2, this occurs when $(x - X)/\sigma = 1.4$ (Figure 5.17). Thus, $(x - X)/\sigma = (760 - 800)/\sigma = 1.4$ and so the standard deviation is 29 MPa. In the above analysis, it was assumed that the mean given was the true value or a good enough approximation.

Example

A series of measurements was made of the periodic time of a simple pendulum and gave a mean of 1.23 s with a standard deviation of 0.01 s. What is the chance that, when a measurement is made, it will lie between 1.23 and 1.24 s?

$(x - X)/\sigma = (1.24 - 1.23)/0.01 = 1.0$. Thus, using Table 5.1, the area between the mean and 1.24 s (Figure 5.18) is 0.3413. Thus about 34% of the readings can be expected to be between 1.23 and 1.24 s. In the above analysis, it was assumed that the mean given was the true value or a good enough approximation.

5.4.1 Significance and confidence levels

With the normal distribution, 95% of the readings fall within 1.96 standard deviations of the mean. Thus, if we obtained a value which differed by more than 1.96 standard deviations from the mean, it could be said to be significantly different from most of the reading values since it would not be one of the 95% and so expected to occur often. In statistical terms we say

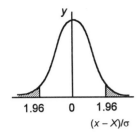

Figure 5.19 *Area for which there is a 5% significance*

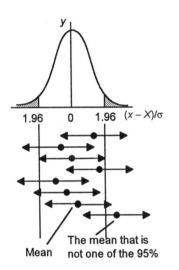

Figure 5.20 *About 95% correctly include the true mean within their error bands*

that a reading which was not within 1.96 standard deviations of the mean is *significant at the 5% level* (Figure 5.19). For a reading to be *significant at the 1% level* its deviation from the mean must be at least 2.576 standard deviations.

Consider a situation in which we have a box in which we think there are five red and five black counters. Suppose we now draw, at random, five counters from the box. If these are all red then we might suspect that the box did not contain five red and five black counters, considering our sample to be a significant result which was highly improbable if there had been five red and five black counters in the box. How small must the probability of this 'improbable' event be for us to doubt the initial hypothesis of five red and five black when it actually occurs? Traditionally the probability, i.e. significance, levels are taken as:

5% and less is deemed significant
1% and less is deemed highly significant
0.1% and less is deemed very highly significant

Thus if an event occurs which has a 5% or less significance level, then the event is regarded as significant. For example, if the value of the acceleration due to gravity in a particular place is expected to be 9.813 m/s² and the mean result of a number of experiments is found to be 9.892 m/s², then if this mean has a 5% significance level we begin to doubt that the expected value of 9.813 m/s² is valid. If the significance level is 1% or less then the result is deemed highly significant and we now have serious doubts about it being 9.813 m/s² and regard it as highly improbable.

Since 95% of the readings in a normal distribution lie within 1.96 standard deviations of the mean X, we can be 95% confident that a reading taken at random will occur within these limits. We thus define what is termed a *95% confidence interval* as:

$$X - 1.96\sigma \leq x \geq X + 1.96\sigma$$

This means that if we consider 95% of the readings, we will be able to describe them as having means that fall within the interval $X \pm 1.96\sigma$. Figure 5.20 illustrates this.

In a similar way we can define a *99% confidence interval* as:

$$X - 2.576\sigma \leq x \geq X + 2.576\sigma$$

99% of the readings occur within 2.576 standard deviations of the mean.

Consider what happens if we take samples of some measurement. As the sample size increases then the standard deviations of the means, i.e. standard error, decreases, being given by $\sigma/\sqrt{n}$. This reduction in standard error means that the confidence interval becomes smaller. This means there is an increase in the precision of the measurement as a result of the increased sample size.

Example

In an experiment, ten measurements of the boiling point of water were made and the following results obtained:

99.8, 100.0, 99.7, 99.9, 99.9, 100.1, 100.1, 99.7, 100.0, 99.9°C

The standard deviation of large numbers of such measurements is known to be 0.2°C. Do the results indicate that the experiment could feasibly, with many more results, give a mean value differing from 100.0°C?

The sample mean is 99.91°C. The standard error of the sample mean is $0.2/\sqrt{10} = 0.063$°C. The value of $(x - X)/\sigma = (99.91 - 100)/0.063$ and is thus 1.42. This is less that 1.96 so the mean of the results is within the 95% confidence interval and so, in the long run, it is feasible that the results could give a mean of 100.0°C.

5.5 t-distribution

It has often been assumed, in earlier examples in this chapter, that the standard deviation of the population is known. Often, however, only the standard deviation of a small sample is known and this has been used as though it were the standard deviation of the population. But this introduces some element of unreliability. Thus if we want to be sure of, say, the 95% confidence level then we need to broaden the interval to allow for it having been calculated from the standard deviation of a small sample. We do this by replacing the values of the areas for $(x - X)/\sigma$ taken from the normal distribution by larger values for a similar distribution, this being called the *Student's t-distribution*. It is called Student because W.S. Gossett, who studied the distribution, published papers about it under the pen name of 'Student'. If we use the sample standard deviation s then, for a sample size n:

$$\text{estimated standard error of the mean } \hat{\sigma} = \frac{s}{\sqrt{n}}$$

and we define t as:

$$t = \frac{x-X}{\hat{\sigma}} = \frac{x-X}{s/\sqrt{n}}$$

When n is more than about 25, the sample standard deviation is close enough to the population standard deviation for $\hat{\sigma}$ to be virtually the same as σ and thus t to be effectively the same as z, i.e. $(x - X)/\sigma$.

When using the t distribution, the 95% confidence interval is given by:

$$X - t_{0.025}\frac{s}{\sqrt{n}} \leq x \geq X + t_{0.025}\frac{s}{\sqrt{n}}$$

The 0.025 subscript to the t indicates that we are taking an ordinate which gives the area fraction 0.025, i.e. 2.5%, from the outer limit of the distribution. For the 99% confidence interval we have:

$$X - t_{0.005} \frac{s}{\sqrt{n}} \leq x \geq X + t_{0.005} \frac{s}{\sqrt{n}}$$

The 0.005 subscript to the t indicates that we are taking an ordinate which gives the area fraction 0.005, i.e. 0.5%, from the outer limit of the distribution.

Table 5.3 gives t values. To use the table you need to know the *degree of freedom* of the data being used. The degree of freedom, usual symbol v, is the denominator used in calculating s, i.e. $n - 1$.

Table 5.3 t critical points

v	$t_{0.25}$	$t_{0.10}$	$t_{0.05}$	$t_{0.025}$	$t_{0.010}$	$t_{0.005}$	$t_{0.0025}$	$t_{0.0010}$	$t_{0.0005}$
1	1.00	3.08	6.31	12.7	31.8	63.7	127	318	637
2	0.82	1.89	2.92	4.30	6.96	9.92	14.1	22.3	31.6
3	0.76	1.64	2.35	3.18	4.54	5.84	7.45	10.2	12.9
4	0.74	1.53	2.13	2.78	3.75	4.60	5.60	7.17	8.61
5	0.73	1.48	2.02	2.57	3.36	4.03	4.77	5.89	6.87
6	0.72	1.44	1.94	2.45	3.14	3.71	4.32	5.21	5.96
7	0.71	1.41	1.89	2.36	3.00	3.50	4.03	4.79	5.41
8	0.71	1.40	1.86	2.31	2.90	3.36	3.83	4.50	5.04
9	0.70	1.38	1.83	2.26	2.82	3.25	3.69	4.30	4.78
10	0.70	1.37	1.81	2.23	2.76	3.17	3.58	4.14	4.59
11	0.70	1.36	1.80	2.20	2.72	3.11	3.50	4.02	4.44
12	0.70	1.36	1.78	2.18	2.68	3.05	3.43	3.93	4.32
13	0.69	1.35	1.77	2.16	2.65	3.01	3.37	3.85	4.22
14	0.69	1.35	1.76	2.14	2.62	2.98	3.33	3.79	4.14
15	0.69	1.34	1.75	2.13	2.60	2.95	3.29	3.73	4.07
16	0.69	1.34	1.75	2.12	2.58	2.92	3.25	3.69	4.01
17	0.69	1.33	1.74	2.11	2.57	2.90	3.22	3.65	3.97
18	0.69	1.33	1.73	2.10	2.55	2.88	3.20	3.61	3.92
19	0.69	1.33	1.73	2.09	2.54	2.86	3.17	3.58	3.88
20	0.68	1.33	1.72	2.09	2.53	2.85	3.15	3.55	3.85
25	0.68	1.32	1.71	2.06	2.49	2.79	3.08	3.45	3.73
30	0.68	1.31	1.70	2.04	2.46	2.75	3.03	3.39	3.65

Example

The chemical analysis of nine different samples of a compound gave the following results for the percentage of a particular element in that compound:

12.2, 10.9, 11.5, 12.1, 11.3, 12.0, 10.8, 12.3, 11.6%

Does the mean of this sample of results differ significantly from the expected value of 12.1%?

The mean of the above data is 11.63 and the standard deviation 0.56. Because the population standard deviation is not known and the sample size is comparatively small, the *t*-distribution is used. For the data:

$$t = \frac{x - X}{s/\sqrt{n}} = \frac{11.63 - 12.1}{0.56/\sqrt{9}} = -2.52$$

The number of degrees of freedom is 9 − 1 = 8. From Table 5.3, $t_{0.025} = 2.31$. Thus, as the $|t|$ value is more than 2.31, the mean result is significant at the 5% level. A level of significance less than 5% is generally taken to be reasonable evidence that a proposition is likely to be untrue. The mean is outside the 95% confidence interval. Thus the percentage might well not be 12.1%.

Example

The results of four measurements of the time for 100 oscillations of a simple pendulum are: 64, 79, 76, 78 s. It is thought that the mean time should be 70 s. Is this likely?

The mean of the above results is 74.25 s and the standard deviation 6.95 s. Thus:

$$t = \frac{x - X}{s/\sqrt{n}} = \frac{74.25 - 70}{6.95/\sqrt{4}} = 1.22$$

The number of degrees of freedom of the data is 4 − 1 = 3. Thus Table 5.3 gives $t_{0.25} = 0.76$ and $t_{0.10} = 1.64$. Thus the result has a confidence level between 50% and 20% and so it is perfectly feasible that the mean time could be 70 s.

Example

According to data tables, the Curie temperature for a particular alloy should be 510°C. In a laboratory experiment values of the temperature were determined for ten samples of the material. The results were:

500, 690, 630, 490, 580, 700, 560, 710, 520, 720°C

Is it likely that the samples used differ in some way from that used to give the value in the tables?

The mean of the results is 610°C and the standard deviation is 91.3°C. Thus:

$$t = \frac{x - X}{s/\sqrt{n}} = \frac{610 - 510}{91.3/\sqrt{10}} = 3.46$$

The number of degrees of freedom is 9. Table 5.3 gives $t_{0.005}$ as 3.25. Thus the result has a confidence level of 1% or less. This is highly significant and it seems likely that the samples used differed in some way from that used for the value given in tables.

5.6 Difference between two means

A problem that sometimes occurs is that the means of two samples of data have to be compared. Thus if the means are $\bar{x}_1$ and $\bar{x}_2$ we need to determine if the difference $\bar{x}_1 - \bar{x}_2$ is significant and so the true values are the same, i.e. $X_1 = X_2$.

5.6.1 Standard error of the population known

When the standard error σ of the population is known and we have n_1 data values for sample 1 and n_2 values for sample 2 taken from that population, then the standard deviations of the two samples are $\sigma/\sqrt{n_1}$ and $\sigma/\sqrt{n_2}$. When we subtract two quantities with errors, the result has an error given by (see section 4.5.2):

$$(\text{standard error of } (\bar{x}_1 - \bar{x}_2))^2 = (\text{error of 1})^2 + (\text{error of 2})^2$$

$$= \frac{\sigma^2}{n_1} + \frac{\sigma^2}{n_2}$$

Thus:

$$\text{standard error of } (\bar{x}_1 - \bar{x}_2) = \sigma\sqrt{\frac{1}{n_1} + \frac{1}{n_2}}$$

Thus we can compute a value for z and determine the level of significance. Since:

$$z = \frac{\text{difference in the two mean}}{\text{standard error}}$$

then, for the difference between the two means:

$$z = \frac{\bar{x}_1 - \bar{x}_2}{\sigma\sqrt{\frac{1}{n_1} + \frac{1}{n_2}}}$$

5.6.2 Standard error of the population not known

Where the standard error of the population is not known we have to use the standard deviations s_1 and s_2 of the samples. The *pooled sample standard deviation* s_p is used as an estimate of the standard error of the population from which the samples are considered to have been taken, being given by:

$$s_p^2 = \frac{(n_1 - 1)s_1^2 + (n_2 - 1)s_2^2}{(n_1 - 1) + (n_2 - 1)}$$

where n_1 is the sample size with standard deviation s_1 and n_2 the sample size with standard deviation s_2. The square of the pooled standard deviation is the weighted average of the squares of the standard deviations of the samples, the weighting according to the number of degrees of freedom of each sample, i.e. $(n_1 - 1)$ and $(n_2 - 1)$. The total number of degrees of freedom of the pooled standard deviation is:

$$\text{degrees of freedom} = (n_1 - 1) + (n_2 - 1) = n_1 + n_2 - 2$$

When the two samples have the same sample size, i.e. $n_1 = n_2$, then the above equation gives:

$$s_p^2 = \frac{s_1^2 + s_2^2}{2}$$

We can then, as in the previous section, use the pooled standard deviation as the value for the standard error and obtain a value for t:

$$t = \frac{\bar{x}_1 - \bar{x}_2}{s_p \sqrt{\frac{1}{n_1} + \frac{1}{n_2}}}$$

The t value thus obtained can be used to determine the significance of the data.

Example

Two samples of resistors, with nominally the same resistance, were sampled from a batch. For the first sample, it was found that when 5 resistors were tested there was a sample mean of 10.7 Ω and a standard deviation of 0.5 Ω. For the second sample, when 4 resistors were taken there was a mean of 10.2 Ω and a standard deviation of 0.7 Ω. Are the means significantly different at the 5% level of significance?

The pooled standard deviation is:

$$s_p^2 = \frac{(n_1 - 1)s_1^2 + (n_2 - 1)s_2^2}{(n_1 - 1) + (n_2 - 1)} = \frac{4 \times 0.5^2 + 3 \times 0.7^2}{4 + 3}$$

and so the pooled standard deviation is 0.594 Ω. Then, using the equation derived above:

$$t = \frac{\bar{x}_1 - \bar{x}_2}{s_p\sqrt{\frac{1}{n_1} + \frac{1}{n_2}}} = \frac{10.7 - 10.2}{0.594\sqrt{\frac{1}{5} + \frac{1}{4}}} = 1.25$$

The total number of degrees of freedom is 4 + 3 = 7. Using Table 5.3, $t_{0.025}$ = 2.36. Thus the difference between the two samples is not significant at the 5% level.

Example

Two students carried out measurements of the percentages of an element in a compound. Student 1 obtained a mean percentage of 3.62 with standard deviation 0.30, and student 2 a mean percentage of 3.25 with standard deviation 0.25. Both used sample sizes of 10. Is there any significant difference between the two results?

Since both the sample sizes are the same, the pooled standard deviation is given by:

$$s_p^2 = \frac{s_1^2 + s_2^2}{2} = \frac{0.30^2 + 0.25^2}{2}$$

and so s_p = 0.276. Thus, using the equation derived above:

$$t = \frac{\bar{x}_1 - \bar{x}_2}{s_p\sqrt{\frac{1}{n_1} + \frac{1}{n_2}}} = \frac{3.62 - 3.25}{0.276\sqrt{\frac{1}{10} + \frac{1}{10}}} = 3.00$$

The number of degrees of freedom is 9 + 9 = 18. Table 5.3 gives $t_{0.005}$ as 2.88. Thus the difference is significant at the 1% level. This would suggest that it is extremely unlikely that the two samples have the same mean.

Problems

1 Determine the mean and the standard deviation for the following data: 10, 20, 30, 40, 50.

2 The following are the results of 100 measurements of the times for 50 oscillations of a simple pendulum:

Between 58.5 and 61.5 s, 2 measurements
Between 61.5 and 64.5 s, 6 measurements
Between 64.5 and 67.5 s, 22 measurements
Between 67.5 and 70.5 s, 32 measurements
Between 70.5 and 73.5 s, 28 measurements
Between 73.5 and 76.5 s, 8 measurements

Between 76.5 and 79.5 s, 2 measurements

(a) Determine the relative frequencies of each segment.
(b) Determine the mean and the standard deviation.

3   The probability that a break could occur in a length of wire 2 m long is the same at all points along the length. Determine the probability that a break will occur between 0.5 m and 1.5 m.

4   Measurements of the resistances of resistors in a batch gave a mean of 12 Ω with a standard deviation of 2 Ω. If the resistances can be assumed to have a normal distribution about this mean, how many from a batch of 300 resistors are likely to have resistances more than 15 Ω?

5   A random sample of 16 measurements are taken from a population which has a mean of 25 and a standard deviation of 6. Determine the 90% confidence interval.

6   The times for 20 oscillations of a simple pendulum give a population of results with a mean time of 19.80 s and a standard deviation of 1.20 s. Determine the 95% confidence interval.

7   A set of six readings of the resistance of a 1 m length of wire gave the results: 18.6, 20.5, 19.2, 18.4, 20.8, 19.4 Ω. Determine the 95% confidence interval for the resistance, assuming that the population is normal.

8   Six measurements were made of the capacitance of a capacitor and the results gave a mean of 14.1 μF. If the standard deviation of repeated measurements is known to be 2.5 μF, determine the 95% confidence interval assuming that the measurements are normally distributed.

9   Data books indicate that the e.m.f. of a thermocouple at a particular temperature should be 12 mV. Measurements gave the values: 12.1, 12.4, 12.3, 11.8, 11.9 mV. Could these measurements be deemed to conform with the data value if the standard deviation of the measurements is 0.4 mV and the population can be assumed to be normal?

10 The length of a wire is claimed to have been set to a nominal value of 1.500 m. When a student measures it the following results are obtained:

   1.502, 1.502, 1.504, 1.501, 1.501, 1.500, 1.500, 1.505,
   1.501, 1.501, 1.503, 1.498, 1.499, 1.502, 1.501, 1.501 m

Do the results suggest that the length has been set to 1.500 m?

11 The time taken for a ball bearing to fall through a particular height is measured and the following results obtained: 12, 28, 17, 15 ms.

Calculation, using the assumed value for the acceleration due to gravity, would suggest that the time should be 25 ms. Is the mean of the measurements significantly different from the assumed value?

12. The accuracy of a voltmeter was tested using a standard cell of e.m.f. 1.10 V and the following results obtained:

    1.11, 1.10, 1.12, 1.14, 1.15, 1.14, 1.09, 1.11, 1.15, 1.13 V

    Do the results suggest that the meter could have a systematic error?

13. Measurements of the concentrations of a particular element in five samples taken from two different batches of an ore gave the results as:

    Batch 1: 22, 12, 11, 16, 19 parts per million
    Batch 2: 8, 16, 12, 9, 10 parts per million

    Determine whether it is likely that the two ore batches could be the same.

14. Measurements were made of the percentage elongation of samples of a material before and after heat treatment and the following results obtained:

    Before: 12.5, 13.9, 14.6, 10.8, 11.0, 11.8 %
    After: 15.4, 13.6, 14.7, 12.1, 13.2, 11.5 %

    Does it seem likely that the treatment changed the percentage elongation?

15. Student A made 10 measurements of the diameter of a rod and obtained the results:

    13, 19, 15, 11, 22, 14, 17, 20, 14, 15 mm

    Student B, on another day, made 5 measurements of what was thought to be the same rod. The results obtained were:

    9, 8, 12, 10, 16 mm

    Determine whether it is likely that they were both using the same rod.

16. Student A made measurements of the resistance of a resistor and obtained the results:

    61, 55, 67, 62, 56, 59 Ω

    Student B, carrying out the same experiment, obtained:

62, 70, 75, 81 Ω

Is the difference between the two sets of results significant?

17 A student carries out measurements of the mass of copper deposited on an electrode in a specified time and obtained the results:

40, 30, 29, 28, 27, 34, 29, 28, 28, 27 mg

The next day the student decides to recheck the results and obtains:

26, 27, 28, 32, 42, 27, 28, 29, 38, 43 mg

Is there any significant difference in the two sets of results?

18 Measurements of the molar heat capacity of a solid at a particular temperature gave the following results:

80.02, 80.02, 80.05, 80.04, 80.03, 80.02, 80.04,
79.98, 80.00, 80.03, 79.97, 80.03, 80.04 J mol^{-1} K^{-1}

Measurements made by another method gave:

79.97, 80.02, 79.94, 79.95, 79.98, 80.03, 79.97, 79.97 J mol^{-1} K^{-1}

Is there any significant difference between the two sets of results?

6 Least squares

In Chapter 3 the problem was considered of determining the best straight line through a series of points on a graph. In this chapter we consider the *least squares method* of fitting a straight line to a plot of points and determining the slope and intercept of the line. In essence, what we have is a method of taking data for experimental points and finding the values of m and c that will give the best fit when that data is inserted into an equation of the form $y = mx + c$.

6.1 The method of least squares

Consider the data points shown in the graph of Figure 6.1 and the best straight line through them. In Chapter 3 the best straight line was drawn by balancing the points so that those above the line balanced those below the line. To simplify the discussion, we will consider that the errors are entirely in the y measurements, i.e. the dependent variable. Thus the best line is drawn by considering the deviations of the points in the y directions from the line and trying to achieve a balance. The deviation d of any one point is given by:

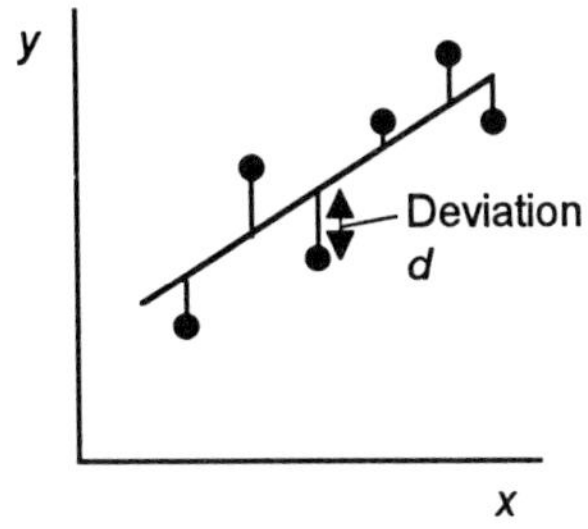

$$\text{deviation } d = y - \hat{y}$$

where y is the measured value and the value given by the straight line. The value given by the straight line is:

$$\hat{y} = mx + c$$

where x is the value of the independent variable, m the gradient of the graph and c its intercept with the y axis. Thus, we can write:

$$d = y - (mx + c)$$

We can write such an equation for each data point. Thus, for point j we can write for its deviation d_j:

$$d_j = y_j - (mx_j + c)$$

with y_j and x_j being the pair of data values.

We might consider that the best line would be the one that minimises the sum of the deviations. However, a better solution is to find the line that minimises the standard errors of the points in their deviations from the straight line. This means choosing the values of m and c which minimise the sum of the squares of these deviations. This is termed the *least squares estimate* for the best straight line. Hence the best values of m and c are obtained:

Figure 6.1 *The best straight line*

sum of the squares $S = \sum_{j=1}^{n} d_j^2 = \sum_{j=1}^{n}[y_j - (mx_j + c)]^2$

We can consider there to be many values of m and c, i.e. lots of straight lines that can be drawn, and we need to determine the values which result in a minimum value for the sum. If we had a simple equation such as $S = mx^2$ we could find the minimum value of the sum with respect to m by differentiating and then equating the derivative to zero, i.e. finding the value of m that gave $dS/m = 0$. But we have two variables, m and c. Thus, in this case we have to determine the partial derivatives $\partial S/\partial m$ and $\partial S/\partial$ and equate them to zero. Expanding the equation for the sum gives:

$$S = \sum \left[y_j^2 - 2y_j(mx_j + c) + (mx_j + c)^2 \right]$$

Hence:

$$\frac{\partial S}{\partial m} = -2 \sum x_j(y_j - mx_j - c)$$

and:

$$\frac{\partial S}{\partial c} = -2 \sum (y_j - mx_j - c)$$

Equating both these partial derivatives to zero gives:

$$-2 \sum x_j(y_j - mx_j - c) = 0$$

$$-2 \sum (y_j - mx_j - c) = 0$$

We can rearrange these equations to give:

$$m \sum x_j^2 + c \sum x_j = \sum x_j y_j \qquad [1]$$

$$m \sum x_j + nc = \sum y_j \qquad [2]$$

Thus all we need to do is put the data into the above two equations and solve the pair of simultaneous equations to obtain values for m and c.

Alternatively, we can solve these two simultaneous equations [1] and [2] to give m as:

$$m = \frac{n \sum x_j y_j - \sum y_j \sum x_j}{n \sum x_j^2 - (\sum x_j)^2} \qquad [3]$$

This form of the equation is a useful form for using with a calculator. We can, however, put the equation in another form. Since the average values of x and y are given by:

$$\bar{x} = \frac{1}{n}\sum x_j$$

$$\bar{y} = \frac{1}{n}\sum y_j$$

we can write m as:

$$m = \frac{\sum x_j(y_j - \bar{y})}{\sum x_j(x_j - \bar{x})}$$

and, since the sum of the deviations of y about the mean value of y must be zero, i.e. $\Sigma(y_j - \bar{y}) = $, then:

$$m = \frac{\Sigma(x_j - \bar{x})(y_j - \bar{y})}{\Sigma(x_j - \bar{x})^2} \quad [4]$$

To obtain a value for c, we can substitute this value of m into equation [2] when rewritten in the form:

$$c = \bar{y} - m\bar{x} \quad [5]$$

Incidentally, this equation is of the form $\bar{y} = m\bar{x} + c$ and so shows that the least squares line passes through the mean values of the data.

To obtain a value for c, an alternative to using equation [5] is to solve the simultaneous equations [1] and [2] for c to obtain:

$$c = \frac{\Sigma x_j^2 \, \Sigma y_j - \Sigma x_j \, \Sigma x_j y_j}{n \Sigma x_j^2 - \left(\Sigma x_j\right)^2} \quad [7]$$

Example

The following data was obtained from an experiment in which the resistance R of a coil of metal wire was determined at different temperatures θ. The relationship between the resistance and temperature is envisaged as being of the form $R = m\theta + c$. Determine the values of m and c which best fit the data.

Resistance (Ω)	76.4	82.7	87.8	94.0	103.5
Temperature (°C)	20.5	32.5	52.0	73.0	96.0

To use equations [1] and [2] we need to find the values of the sum of the temperature data, the sum of the squares of the temperature data, the sum of the resistance data and the sum of the products of the two sets of data. Table 6.1 shows the calculations of these values.

Table 6.1 *Calculations for example*

R (Ω)	θ (°C)	θ^2 (°C)^2	$R\theta$ (Ω °C)
76.4	20.5	420.25	1566.20
82.7	32.5	1056.25	1687.75
87.8	52.0	2704.00	4565.60
94.0	73.0	5329.00	6862.00
103.5	96.0	9216.00	9936.00
$\Sigma R = 444.4$	$\Sigma \theta = 274$	$\Sigma \theta^2 = 18\,725.50$	$\Sigma R\theta = 25\,617.55$

Hence equation [1], namely:

$$m \sum x_j^2 + c \sum x_j = \sum x_j y_j$$

becomes:

$$18\,725.50m + 274c = 25\,617.55$$

Equation [2], namely:

$$m \sum x_j + nc = \sum y_j$$

becomes:

$$274m + 5c = 444.4$$

These two simultaneous equations can then be solved to give the values of m and c.

Multiplying the first equation by 5/274 gives:

$$341.71m + 5c = 467.47$$

Subtracting the second equation from the above equation gives:

$$67.71m = 23.07$$

Hence $m = 0.341$. Substituting this value into one of the equations gives $c = 70.19$. Hence the equation is:

$$R = 0.341\theta + 70.19$$

As an alternative to putting values in the simultaneous equations [1] and [2], we could have used equations [3] and [2] or equations [4] and [5]. Using equation [3], namely:

$$m = \frac{n\Sigma x_j y_j - \Sigma y_j \, \Sigma x_j}{n\Sigma x_j^2 - (\Sigma x_j)^2}$$

and taking the values from Table 6.1:

$$m = \frac{5 \times 25\ 617.55 - 444.4 \times 274}{5 \times 18\ 725.50 - 274^2}$$

and hence, as before, $m = 0.341$. c can be found as before.

To use equations [4] and [5], we need to tabulate the data in a different form. Table 6.2 shows the table. Using equation [4], namely:

$$m = \frac{\Sigma(x_j - \bar{x})(y_j - \bar{y})}{\Sigma(x_j - \bar{x})^2}$$

gives:

$$m = \frac{1263.59}{3710.30} = 0.341$$

Using equation [5], namely:

$$c = \bar{y} - m\bar{x}$$

gives:

$$c = 88.88 - 0.341 \times 54.8 = 70.19$$

Thus, as before:

$$R = 0.341\theta + 70.19$$

Table 6.2 *Calculations for example*

$R\ (\Omega)$	$\theta\ (°C)$	R − mean (Ω)	θ − mean $(°C)$	$(R$ − mean$) \times (\theta$ − mean$)$	$(\theta$ − mean$)^2$
76.4	20.5	−12.48	−34.3	428.06	1176.49
82.7	32.5	−6.18	−22.3	137.81	497.29
87.8	52.0	−1.08	−2.8	3.02	7.84
94.0	73.0	5.12	18.2	93.18	331.24
103.5	96.0	14.6	41.2	601.52	1697.44
Mean = 88.88	Mean = 54.8			$\Sigma = 1263.59$	$\Sigma = 3710.30$

Example

The following is data obtained for the volume of a gas measured when it is measured at different volumes. The relationships between the volume v and pressure p is expected to be of the form $pv^\gamma = C$, where γ and C are constants. Assuming that the only significant errors are in the volume, determine a value for γ.

p (10^5 Pa)	0.5	1.0	1.5	2.0	2.5	3.0
v (dm^3)	1.62	1.00	0.75	0.62	0.52	0.46

Taking logarithms of the equation gives:

$$\lg p + \gamma \lg v = \lg C$$

If we let $X = \lg p$ and $y = \lg v$ then, with some rearrangement, we have:

$$Y = (-1/\gamma) X + (1/\gamma) \lg C = mX + c$$

We can use the least squares method with the data and equations [3] and [2]. Table 6.3 giving intermediate calculation.

Table 6.3 *Calculations for example*

lg v, i.e. Y	lg p, i.e. X	XY	X^2
−0.3010	0.2095	−0.0631	0.0906
0	0	0	0
0.1761	−0.1249	−0.0220	0.0310
0.3010	−0.2076	−0.0625	0.0906
0.3939	−0.2840	−0.1130	0.1583
0.4771	−0.3392	−0.1609	0.2276
$\Sigma = 1.0511$	$\Sigma = -0.7442$	$\Sigma = -0.4215$	$\Sigma = 0.5981$

Equation [4] gives

$$m = \frac{n \Sigma x_j y_j - \Sigma y_j \Sigma x_j}{n \Sigma x_j^2 - (\Sigma x_j)^2}$$

$$= \frac{6 \times (-0.4215) - 1.0511 \times (-0.7442)}{6 \times 0.5981 - 1.0511^2} = -0.7033$$

Substituting this value in equation [2] gives:

$$m \sum x_j + nc = \sum y_j$$

$$-0.7033 \times 1.0511 + 6c = -0.7442$$

Hence $c = -0.000827$ and so the equation is:

$$\lg v = -0.7033 \lg p - 0.000827$$

Thus $(-1/\gamma) = -0.7033$ and so $\gamma = 1.42$.

6.1.1 Using a calculator

The form outlined in the above example is convenient for use with a calculator. However, some calculators have a least squares facility which enables data to be entered directly into the calculator and an output of the gradient and intercept obtained. With the Casio calculators with this facility, the sequence of key operations is as follows:

1. Put the calculator into the least squares mode by pressing the MODE 2 key. The letters LR (standing for linear regression) appear in the display.

2. To enter the first x data point, key in the value followed by the [(--- key.

3. Then enter the corresponding y data point, key in the value followed by the M+ key.

4. Repeat operations 2 and 3 for each pair of data values.

5. To obtain the gradient, press Shift and the 8 key. Casio use the symbol B for the gradient.

6. To obtain the intercept, press Shift and the 7 key. Casio use the symbol A for the intercept.

6.2 Errors

In the calculations of the gradient and intercept given above, no estimate was given of the possible error in their values. We used values for y with no assumption of any error in the value used. But we might repeat the measurement of the y value and obtain a different value for the same x value. Indeed we might obtain a distribution of y values for a particular x value.

For example, in an experiment involving a simple pendulum, we might measure the time taken for 20 oscillations when the pendulum length is 0.50 m and repeat the time measurement at that length 5 times. We can thus obtain the error in the time, the y value, at a fixed value of length x. We then obtain another data point by repeating the time measurement for a length of 0.60 m. If we give equal weight to each data point then we are assuming that each point has the same standard error in its y values, i.e. the standard error in the time quoted for each of these two lengths is the same. In this section we will make such a simplifying assumption. Later in this chapter we will consider the situation when the errors are not the same.

We can thus consider that we have effectively a set of y values at each x value. For each y value at each x value in combination with each of the values at the other x values, we can obtain, by the least squares method, values of the gradient and intercept. Thus we end up with an entire set of straight lines distributed about the mean straight line. What we need to find out is the standard error of the gradients and of the intercepts.

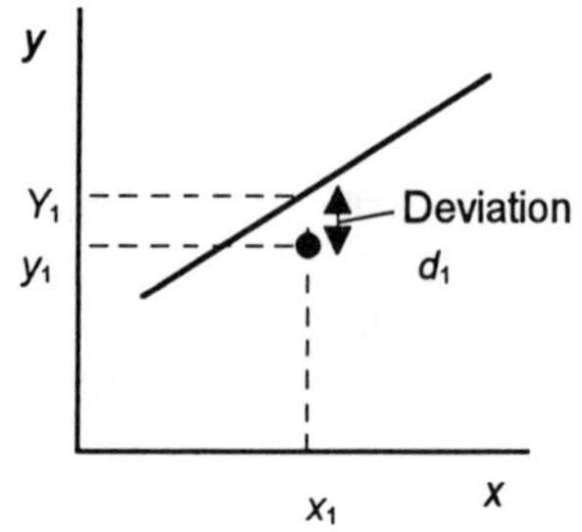

The deviation d_1 of a point at y_1 from the straight line (Figure 6.2), where Y is the value on the line at x_1, is:

$$d_1 = y_1 - Y_1 = y_1 - mx_1 - c$$

We can write similar equations for each point. All the deviations of the measured points about the mean straight line are typical of the deviation about that line that could occur with just one of the points. The standard deviation σ of these points is thus:

$$\sigma^2 = \frac{d_1^2 + d_2^2 + d_3^2 + \ldots d_n^2}{n-2} = \frac{1}{n-2} \sum d_j^2$$

We are dividing by $(n-2)$, where n is the number of data points, because there are two restrictions placed on the possible values of y because if we only had two points there would be only one possible line and it would necessarily be the best line. Thus the degree of freedom is reduced by 2 when we consider possible straight lines.

This uncertainty in the value of y will contribute to an uncertainty in the values of m and c. The gradient is given by equation [4] as:

$$m = \frac{\Sigma(x_j - \bar{x})(y_j - \bar{y})}{\Sigma(x_j - \bar{x})^2}$$

If M is the true value of the gradient given by the values of Y, then:

$$M = \frac{\Sigma(x_j - \bar{x})(Y_j - \bar{y})}{\Sigma(x_j - \bar{x})^2}$$

Hence:

$$m - M = \frac{\Sigma(x_j - \bar{x})(y_j - \bar{y})}{\Sigma(x_j - \bar{x})^2} - \frac{\Sigma(x_j - \bar{x})(Y_j - \bar{y})}{\Sigma(x_j - \bar{x})^2}$$

$$= \frac{\Sigma(x_j - \bar{x})(y_j - Y_j)}{\Sigma(x_j - \bar{x})^2}$$

Figure 6.2 *Deviation*

$(m - M)$ is the deviation of the gradient from its true line and $(y_j - Y_j)$ the deviation of the y values from their true values, hence the standard deviation σ_m of the gradient is given by:

$$\sigma_m^2 = \frac{\sigma^2 \Sigma(x_j - \bar{x})^2}{\left[\Sigma(x_j - \bar{x})^2\right]^2} = \frac{\sigma^2}{\Sigma(x_j - \bar{x})^2}$$

Since $\Sigma(x_j - \bar{x})^2 = \Sigma x_j^2 - n\bar{x}^2$, we can write the above equation as:

$$\sigma_m^2 = \frac{\sigma^2}{\Sigma x_j^2 - n\bar{x}^2}$$

and so:

$$\sigma_m^2 = \frac{\Sigma d_j^2}{(n-2)\left(\Sigma x_j^2 - n\bar{x}^2\right)}$$

Similarly the standard deviation σ_c for the intercept values can be derived from the simultaneous equations as:

$$\sigma_c^2 = \frac{\Sigma d_j^2 \Sigma x_j^2}{n(n-2)\left(\Sigma x_j^2 - n\bar{x}^2\right)}$$

Since $\bar{x} = \frac{1}{n}\Sigma x_j$, we can also write the equations as:

$$\sigma_m^2 = \frac{\Sigma d_j^2}{(n-2)\left[\Sigma x_j^2 - \frac{1}{n}\left(\Sigma x_j\right)^2\right]} = \frac{n\Sigma d_j^2}{(n-2)\left[n\Sigma x_j^2 - \left(\Sigma x_j\right)^2\right]}$$

and:

$$\sigma_c^2 = \frac{\Sigma d_j^2 \Sigma x_j^2}{n(n-2)\left[\Sigma x_j^2 - \frac{1}{n}\left(\Sigma x_j\right)^2\right]} = \frac{\Sigma d_j^2 \Sigma x_j^2}{(n-2)\left[n\Sigma x_j^2 - \left(\Sigma x_j\right)^2\right]}$$

There is an alternative way of deriving the standard error for the intercept and that uses equation [5], namely:

$$c = \bar{y} - m\bar{x}$$

The error in c arises from errors in two terms, namely $\bar{y}$ and m, the error in x and hence $\bar{x}$ having been assumed to be insignificant. In section 4.5.2, when two quantities with standard errors where added or subtracted, the square of the resulting standard error was equal to the sums of the squares of the two errors. Thus:

$$\sigma_c^2 = \sigma_{\bar{y}}^2 + (\sigma_m \bar{x})^2$$

$\sigma_{\bar{y}}$ is the standard error of the mean of y and thus:

$$\sigma_c^2 = \frac{\Sigma d_j^2}{n(n-2)} + (\sigma_m \bar{x})^2$$

This is, with some manipulation, the same as the equation for σ_c derived from the simultaneous equations.

Example

Determine the standard errors for the example given earlier in this chapter which gave the straight line $R = 0.341\theta + 70.19$.

Taking the data values given in that example we can derive the deviations of the points from the line. Table 6.4 shows the result.

Table 6.4 *Derivation of the deviations*

R (Ω)	θ (°C)	True R (Ω)	Deviation (Ω)	(Deviation)^2 (Ω^2)
76.4	20.5	77.1805	−0.7805	0.6092
82.7	32.5	81.2725	1.4275	2.0378
87.8	52.0	87.9220	−0.1220	0.0149
94.0	73.0	95.0830	−0.1083	0.0173
103.5	96.0	102.926	0.5740	0.3295
				$\Sigma = 3.0087$

Thus:

$$\sigma_c^2 = \frac{n \Sigma d_j^2}{(n-2)\left[n \Sigma x_j^2 - \left(\Sigma x_j\right)^2\right]} = \frac{5 \times 3.0087}{3(5 \times 18\,725.50 - 274^2)}$$

and so the standard error in the gradient is 0.016. the standard error in the intercept is given by:

$$\sigma_c^2 = \frac{\Sigma d_j^2 \Sigma x_j^2}{(n-2)\left[n \Sigma x_j^2 - \left(\Sigma x_j\right)^2\right]} = \frac{3.0087 \times 18\,725.50}{3(5 \times 18\,725.50 - 274^2)}$$

and hence is 1.0. Alternatively, we could calculate σ_c by using:

$$\sigma_c^2 = \frac{\Sigma d_j^2}{n(n-2)} + \left(\sigma_c \bar{x}\right)^2 = \frac{3.0087}{5 \times 3} + 0.016^2 \times 54.8^2$$

and this gives the standard error as 1.0. Thus the equation can be written as:

$$R = (0.34 \pm 0.02)\theta + (70 \pm 1)$$

6.3 Weighting of results

In the method of least squares discussed in sections 6.1 and 6.2, the data used for each point was assumed to have the same standard error and so was given equal weighting. However, in some situation the errors ascribed to individual points may differ and so this assumption cannot be made. This might occur because each data point is the result of multiple measurements with, as a consequence, different standard deviations of the data values. It might occur because there is a change in range used with an instrument and the reading error changes. As a consequence, in determining the best straight line, more importance has to be attached to the most precise points. This means that the sum of the squares has to be weighted so that when a line is fitted to the points, the calculated line lies closest to the most precise points.

6.3.1 The mean value and its error

Consider a situation where student A make 7 measurements $z_1, z_2, z_3, \ldots z_7$. The mean value x_1 of the set of results is:

$$x_1 = \tfrac{1}{7}(z_1 + z_2 + z_3 + \ldots + z_7)$$

Suppose student B makes 3 measurements z_8, z_9 and z_{10}. The mean value x_2 of this set of results is:

$$x_2 = \tfrac{1}{3}(z_8 + z_9 + z_{10})$$

If we take into account the number of measurements made by each students, then the best value $\bar{x}$ is not the average of x_1 and x_2 but:

$$\bar{x} = \tfrac{1}{10}(z_1 + z_2 + z_3 + \ldots + z_{10})$$

We can write this as:

$$\bar{x} = \frac{7x_1 + 3x_2}{10}$$

The 7 and the 3 represent *weighting factors* or *relative weights* which we use on the two sets of data. The 10 is the sum of all the weights. Thus, if we have n data values $x_1, x_2, x_3, \ldots x_n$ with relative weights $w_1, w_2, w_3, \ldots w_n$, then the best value $\bar{x}$ will be given by:

$$\bar{x} = \frac{\sum w_j x_j}{\sum w_j}$$

Now consider the situation where we have n measurements of the quantity x each having its own standard error. Thus we have $x_1 \pm \Delta x_1$, $x_2 \pm \Delta x_2$, $x_3 \pm \Delta x_3$, ... $x_n \pm \Delta x_n$ (the symbol Δx has been used instead of the usual symbol of σ to avoid a multiplicity of subscripts in distinguishing the standard error of individual measurements from the overall standard error). As in the simple example with the 7 and the 3 measurements, we suppose that all the measurements were made from the same distribution but some were the average of more samples and so have to be weighted accordingly. The sum of all the weights is, as before, the sum of all the values in the distribution from which the samples were taken. The weighting factor w_1 for x_1 is n_1 where n_1 is the number of values considered to have been summed to give the value z_1, i.e. $w_1 = n_1$. Thus the standard error for x_1 is:

$$\Delta x_1 = \frac{\sigma}{\sqrt{n_1}}$$

where σ is the standard error of the distribution from which each of the samples has been taken. Thus we can write:

$$w_1 = \frac{\sigma^2}{(\Delta x_1)^2}$$

Hence, in general:

$$w_j = \frac{\sigma^2}{(\Delta x_j)^2}$$

Note that the weighting for a measurement is inversely proportional to the square of its standard error. The standard error in $\bar{x}$ is thus:

$$\Delta \bar{x} = \frac{\sigma}{\sqrt{\sum n_j}}$$

Hence, we can write for the best value for x:

$$\bar{x} = \frac{\sum w_j x_j}{\sum w_j} = \frac{\sum \left(\frac{\sigma}{\Delta x_j}\right)^2 x_j}{\sum \left(\frac{\sigma}{\Delta x_j}\right)^2} = \frac{\sum \left(\frac{1}{\Delta x_j}\right)^2 x_j}{\sum \left(\frac{1}{\Delta x_j}\right)^2}$$

and its standard error:

$$\Delta \bar{x} = \frac{\sigma}{\sqrt{\sum n_j}} = \frac{\sigma}{\sqrt{\sum \left(\frac{\sigma}{\Delta x_j}\right)^2}} = \frac{1}{\sqrt{\sum \left(\frac{1}{\Delta x_j}\right)^2}}$$

We can write this as:

$$\frac{1}{(\Delta \bar{x})^2} = \sum \frac{1}{(\Delta x_j)^2}$$

Thus if, for example, we take the results of two measurements of some quantity with each having a different standard error $x_1 \pm \Delta x_1$ and $x_2 \pm \Delta x_2$, the weighted best mean of the two results is given by:

$$\text{best mean} = \left(\frac{1}{\frac{1}{\Delta x_1^2} + \frac{1}{\Delta x_2^2}} \right) \left(\frac{x_1}{\Delta x_1^2} + \frac{x_2}{\Delta x_2^2} \right)$$

and its standard error by:

$$\frac{1}{(\text{standard error})} = \frac{1}{\Delta x_1^2} + \frac{1}{\Delta x_2^2}$$

Example

Determine the weighted mean and its standard error of the following measurements (Table 6.5) of a quantity measured by six different scientists:

Table 6.5 *Measurement values*

Measured value	Standard error
169.0	8.0
165.8	1.8
171.0	10.0
168.6	2.0
167.5	0.9
166.7	2.4

Each value is weighted as the inverse of the square of its standard deviation. Table 6.6 shows the results and the steps in the calculation for the mean value and the standard error.

Table 6.6 *Steps in the calculations*

x	Δx	w	wx
169.0	8.0	0.016	2.70
165.8	1.8	0.309	51.23
171.0	10.0	0.010	1.71
168.6	2.0	0.250	42.15
167.5	0.9	1.235	206.86
166.7	2.4	0.174	29.01
$\Sigma = 1008.6$		$\Sigma = 1.994$	$\Sigma = 333.66$

Hence the weighted mean value is:

$$\bar{x} = \frac{\Sigma w_j x_j}{\Sigma w_j} = \frac{333.66}{1.994} = 167.3$$

The standard error in the mean can be determined using:

$$\Delta \bar{x} = \frac{1}{\sqrt{\Sigma \left(\frac{1}{\Delta x_j}\right)^2}}$$

But w_j is proportional to $(1/\Delta x_j)^2$ and so the sum of all the weights is proportional to the sum of all the $(1/\Delta x_j)^2$ terms, i.e. 1.994. Thus, since we have an error of 2.4 proportional to a $(1/\Delta x_j)^2$ term of 0.174 (the last row in Table 6.6), then:

$$\Delta \bar{x} = \sqrt{\frac{0.174}{1.994}} \times 2.4 = 0.71$$

Thus the result can thus be quoted as 167.3 ± 0.7.

6.3.2 The best line

Consider the problem of determining the equation of the best line through a set of data points (x_1, y_1), (x_2, y_2), (x_2, y_3) ... (x_j, y_j). With points of equal weight we had (section 6.1) for the sum of the squares of the deviations:

$$S = \sum_{j=1}^{n} d_j^2 = \sum_{j=1}^{n} \left[y_j - (mx_j + c) \right]^2$$

If, however, the points have weights of $w_1, w_2, w_3, \ldots w_j$, then the sum of the weighted squares is:

$$S = \sum_{j=1}^{n} w_j d_j^2 = \sum_{j=1}^{n} w_j \left[y_j - (mx_j + c) \right]^2$$

Note that, as derived in section 6.3.1, the weighting for a measurement is inversely proportional to the square of its standard error. Determining the minimum value of S as earlier, equations [1] and [2] become:

$$m \sum w_j x_j^2 + c \sum w_j x_j = \sum w_j x_j y_j$$

$$m \sum w_j x_j + nc = \sum w_j y_j$$

and we can then obtain:

$$m = \frac{\sum w_j x_j y_j - \frac{1}{\sum w_j} \sum w_j x_j \sum w_j y_j}{\sum w_j x_j y_j - \frac{1}{\sum w_j} \left(\sum w_j x_j \right)^2}$$

which is frequently written in the form:

$$m = \frac{\sum w_j x_j y_j - \frac{1}{\sum w_j} \sum w_j x_j \sum w_j y_j}{D}$$

where:

$$D = \sum w_j x_j y_j - \frac{1}{\sum w_j} \left(\sum w_j x_j \right)^2$$

Often a more useful form for calculation is obtained if we replace the weighting factors by the reciprocals of the squares of the standard errors, then:

$$m = \frac{\sum \frac{1}{\sigma_j^2} \sum \frac{x_j y_j}{\Delta x_j^2} - \sum \frac{x_j^2}{\sigma_j^2} \sum \frac{y_j^2}{\sigma_j^2}}{\sum \frac{1}{\sigma_j^2} \sum \frac{x_j y_j}{\sigma_j^2} - \left(\sum \frac{x_j}{\sigma_j^2} \right)^2}$$

The intercept c is given by:

$$c = \bar{y} - m\bar{x}$$

with:

$$\bar{x} = \frac{\Sigma w_j x_j}{\Sigma w_j} = \frac{\Sigma\left(\frac{x_j}{\sigma_j^2}\right)}{\Sigma\left(\frac{1}{\sigma_j^2}\right)}$$

$$\bar{y} = \frac{\Sigma w_j y_j}{\Sigma w_j} = \frac{\Sigma\left(\frac{y_j}{\sigma_j^2}\right)}{\Sigma\left(\frac{1}{\sigma_j^2}\right)}$$

We can derive the standard errors in the same way as used for the unweighted points. The standard error σ_m in the gradient is then given by:

$$\sigma_m^2 = \frac{1}{\Sigma\left(\frac{x_j - \bar{x}}{\sigma_j^2}\right)}$$

Using $c = \bar{y} - m\bar{x}$, the square of the standard error σ_c is the sum of the square of the standard error in $\bar{y}$ and the square of that in $m\bar{x}$ (see section 4.5.2 for the subtraction of two quantities having errors), hence the standard error σ_c is given by:

$$\sigma_c^2 = \frac{1}{\Sigma\left(\frac{1}{\sigma_j^2}\right)} + (\bar{x}\sigma_m)^2$$

Problems

1. Determine, by means of the least squares method and assuming that the only significant errors are in the y values, the best straight line relationship to fit the following data:

x	65	68	71	75	77	80	84	87	93	98
y	72	72	80	82	74	78	89	91	96	95

2. The extension e of a spring is measured when different loads W are applied to stretch it and the following data obtained:

W (kg)	1.0	2.0	3.0	4.0	5.0
e (mm)	8	16	27	32	35

If the error is only in the extension values, determine the gradient and the intercept of the best straight line through the points.

3 The mass m of a compound which will dissolve in 100 g of water is measured at different temperatures θ and the following results obtained:

θ (°C)	0	10	20	30	40	50	60	70	80	90	100
m (g)	43.5	49.5	55.2	60.6	65.5	70.2	75.5	80.0	85.0	89.2	94.0

If the error is only in the mass values, determine the gradient and the intercept of the best straight line through the points.

4 The depression d of the end of a cantilever as a result of weight W being added was measured and the following results obtained:

W (kg)	0.25	1.00	2.25	4.00	6.25
d (mm)	2.2	10.0	22.3	39.4	61.7

If the error is only in the depression values, determine the gradient and the intercept of the best straight line through the points.

5 Determine, by means of the least squares method and assuming that the only significant errors are in the y values, the best straight line relationship to fit the following data:

x	1	2	3	4	5	6
y	1	2	2	3	5	5

6 Determine, by means of the least squares method and assuming that the only significant errors are in the y values, the best straight line relationship to fit the following data:

x	1	2	3	4	5	6	7	8	9	10
y	2.3	4.5	5.2	6.0	8.0	9.8	11.1	13.5	14.0	16.6

7 Measurements are made of the effort E to raise load W and the following results obtained:

W (N)	14	42	84	112
E (N)	6.1	14.3	27.0	36.3

Determine, assuming the only significant errors are in E, the equation of the best straight line through the data.

8 The following data is expected to fit a relationship of the form $y = b\,e^{ax}$. By putting the relationship in a form which would give a straight line graph, use the least squares method to determine the values of a and b. The errors are only significant for y.

x	1.00	1.25	1.50	1.75	2.00
y	5.10	5.79	6.53	7.45	8.46

9   The following data is expected to fit a relationship of the form $y = ax^n$. By putting the relationship in a form which would give a straight line graph, use the least squares method to determine the values of a and n. The errors are only significant for y.

x	5.7	11.1	23.4	31.5	42.7
y	7.71	23.95	85.06	141.0	236.9

10 The following data is expected to fit a relationship of the form $y = a\,e^{kx}$. By putting the relationship in a form which would give a straight line graph, use the least squares method to determine the values of a and n. The errors are only significant for y.

x	0	1	2	3
y	1.05	2.10	3.85	8.30

11 Three students measured the resistance of a resistor and obtained the results:

Student A: $20.105 \pm 0.003\ \Omega$
Student B: $20.165 \pm 0.004\ \Omega$
Student C: $20.130 \pm 0.002\ \Omega$

Determine the combined weighted estimate of the resistance and its error.

12 The results of two methods of determining the acceleration due to gravity gave the results:

Method A: 9.8106 ± 0.0006 m/s^2
Method B: 9.8110 ± 0.0010 m/s^2

Determine the combined weighted estimate and its error.

7 Spreadsheets

The data analysis techniques discussed so far in this book have involved pencil and paper calculations or the use of a pocket calculator. This chapter indicates how *spreadsheets* can be used. As the name 'spreadsheet' suggests, it is simply a means of spreading figures over a sheet of paper and carrying out calculations on them. Though spreadsheets have traditionally been used by accountants for preparing financial accounts, they are now used for a wide range of jobs involving the manipulation of data.

Software, such as the general mathematical programs Mathcad and Mathematica or specialist statistical programs such as SPSS, SYSTAT and UNISTAT, can also be used to carry out analysis of experimental data and plot graphs; the reader is, however, referred to the relevant manuals for details of their use. This chapter is restricted to the basic elements of speadsheets.

7.1 Spreadsheets   The term *spreadsheet* is used for a computer program that organises data in the form of a table and in which the various entries are related to each other. Essentially a spreadsheet consists of a large number of boxes, or *cells* as they are commonly termed, arranged in vertical columns and horizontal rows. Each cell in the table has a unique address. Columns are generally identified by the letters of the alphabet A, B, C, etc., with after 26 columns AA, AB, AC, etc., and rows by numbers 1, 2, 3, etc. For example, Microsoft Excel has a possible 256 columns and 16 384 rows. Figure 7.1 shows the start of such a table with the addresses of some cells. Cell addresses are always specified first by the letter of the column and then the number of the row. Thus, for example, B2 refers to the cell in the B column and row 2. A spreadsheet is thus just a table of cells with unique addresses for each cell.

	A	B	C	D	E	F	etc.
1	A1	B1	C1	D1	E1	F1	
2	A2	B2	C2	D2	E2	F2	
3	A3	B3	C3	D3	E3	F3	
4	A4	B4	C4	D4	E4	F4	
5	A5	B5	C5	D5	E5	F5	
6	A6	B6	C6	D6	E6	F6	
etc.							

Figure 7.1 *Cell addresses*

116 *Experimental Methods*

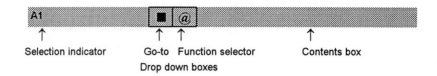

Figure 7.2 *An example of an edit line*

An *edit line* (Figure 7.2) or *control panel* typically appears at the top of the screen, the form depending on the particular spreadsheet used. The edit line gives the cell address which has been selected by moving the cursor to it using the arrow keys or a mouse, the cell being highlighted on the display screen (Figure 7.3). Drop down menus can be produced by selecting the Go-to or navigator box and the function selector. The Go-to menu enables movement to a specific area of the spreadsheet to occur and the function selector menu allows selection of functions for use in formulas (see later). The final box displays the contents of a particular cell when it is selected. At the bottom of the screen there will generally be a status bar which gives information about the current selection of number format, i.e. number of decimal places, etc., the font being used and its size, the date and the mode indicator which displays the current mode of operation. Thus when the spreadsheet is ready to accept data or commands it reads READY and when the command menu is invoked it displays MENU.

Figure 7.3 *Selection of address B4*

The data that may be entered into a cell may be an item of text, a numerical value or a formula involving numerical values held in other cells. Note that when entering numbers, commas should not be used for separating thousands or as a decimal point. Figures containing decimal places must be entered using a full stop for the decimal point. Thus fifty thousand should be entered as 50000 and not 50,000; five point three should be entered as 5.3. The spreadsheet is likely to have a default setting for the number of decimal places that it is set up for and will automatically shorten numbers to that number of places. Thus if the software has a default setting of two decimal places and you enter 1.234567, then the number will be automatically shortened to 1.23.

The number of cells available extends beyond those that can be seen on the screen. To move from cell to cell the arrow keys ←↑→↓ or the mouse might be used. To move from page to page horizontally, Ctrl→ and Crtl← or Scroll Lock Ctrl→ and Crtl← might be used. To move from page to page vertically PgDn and PgUp is used. To move to a specific cell a Go to prompt might be selected by using a mouse to select the facility from a pull down menu, alternatively a Go to key might be used. This is generally the function key F5. When this key is pressed the program prompts you for the address which you can then type in. After pressing Enter the cursor then moves to that cell.

7.1.1 Entering data, text and formulas in cells

Data is entered by the cell being selected, by clicking the mouse when the cursor has been moved on to the cell or using arrow keys, and then the data just directly typed into the space. The other way of entering data into a cell is to use the edit line. As before the cell is selected. The mouse is then clicked in the edit line and the data then typed into the space in that line. Whichever way is used, the data is displayed in both the cell and the edit line.

Text can be included in cells as labels to help the reader understand what columns and rows signify. It is entered by selecting the cell and then typing the text. When the text entered into a cell is more than the width of the cell it spills over into the adjoining cell or cells, assuming they already contain no values. Figure 7.4 illustrates this. The label 'Acceleration due to gravity (m/s^2)' spans A1, B1, C1 and D1 with its value to be put into cell E1. The entries in cells A2, B2 and C2 are intended to be like table headings and signify what values are to be put in the columns they head.

	A	B	C	D	E	F	etc.
1	Acceleration due to gravity (m/s^2)						
2	d (m)	t (s)	d/t (m/s)				
3							
4							
5							
6							
etc.							

Figure 7.4 *Text in a spreadsheet*

A formula is entered into a cell by selecting that cell and then typing the formula in the cell, pressing Enter when complete. The edit line will display the formula, the cell itself will just show the result of using the formula. In order that the entry can be recognised as a formula and not a value or text, it is usually preceded by some special character such as a + or =. The = sign is probably the easiest to remember to use, since the software is being instructed that the value to be put into a particular cell is equal to that given by the formula that follows.

7.1.2 Formulas for relationships between cells

The relationships between the various cells can be defined by a formula so that when data is entered into some cells, the formula can be used to manipulate that data and present the results of the calculations in other cells. The standard operators used with numerical formulas are:

+ Addition / Division
- Subtraction ^ Exponentiation, i.e. to the power
* Multiplication

Note that when writing formulas, do *not* put spaces between characters.
For the addition of the data in cell B1 to that in cell B2 we might have:

=B1+B2 *or* +B1+B2

For subtraction of, say, B2 from B1 we might have:

=B2-B5 *or* +B2-B5

For multiplication of, say, B1 and B2 we use the * sign to indicate multiplication and might have:

=B2*B5 *or* +B2*B5

When data from a cell is to be raised to a power, the sign ^ is generally used to indicate that the number that follows is the power. Thus if we want to square the value in cell B2 we might write:

=B2^2 *or* +B2^2

If we want to divide the data in cell B2 by that in B1 we use the / sign and might write:

=B2/B1 *or* +B2/B1

Brackets can be used in formulas to indicate an order of operations. For example:

=(10*B2)–B1 *or* +(10*B2)–B1

means that the data at B2 is multiplied by 10 and then has the data at B1 subtracted from it.

With some formulas it is possible to use some predefined formulas that are available with the software. In Lotus 1-2-3 these are termed @ functions and are specified by letters preceded by the @ sign. For example, to carry out a summation of the entries in cells B1, B2, B3, B4 and B5, we might have:

=B1+B2+B3+B4+B5 *or* +B1+B2+B3+B4+B5

or use the @ function for summation, namely SUM and write:

=@SUM(B1..B5) *or* +@SUM(B1..B5)

The .., or sometimes :, is used to indicate all the addresses in the column between B1 and B5. The @ functions can be entered by directly typing them in, or using the @ function selector in the edit line, clicking on it and then on the requisite function from the list that drops down, or using the @ function selector and then clicking on the List All in the drop down list. This then presents a much greater list than the shortened one that otherwise appears.

The following are just a few of the functions that are likely to be available.

@AVG	Calculates the average or mean value of a range of cells, e.g. @AVG(B1:B9).
@COUNT	Counts the number of non-blank cells in a range, e.g. @COUNT(A1:A5).
@EXP	Calculates e to the power specified, e.g. @EXP(2) is e^2.
@FACT	Calculates the factorial of the number specified, e.g. @FACT(5) is $5 \times 4 \times 3 \times 2 \times 1$.
@LN	Calculates the natural logarithm, i.e. to base e, of the number that follows or is specified by the cell address, e.g. @LN(2).
@LOG	Calculates the common logarithm, i.e. to base 10, of the number that follows or is specified by the cell address, e.g. @LOG(2).
@MAX	Finds the largest value in the selected range.

@MEDIAN	Finds the median value in the selected range.
@MIN	Finds the minimum value in the selected range.
@PI	Inserts the value for the constant π.
@RAND	Generates a random number between zero and one.
@ROUND	This rounds the value to the nearest power of ten specified, e.g. @ROUND(B2;0) rounds the content of cell B2 to 10^0, i.e. 1, and so the nearest integer.
@SMALL	This finds the smallest value in a specified range.
@SQRT	This gives the square root of the value in the specified cell, e.g. @SQRT(B2).
@STD	This calculates the standard deviation of a specified range, e.g. @STD(B1:B6).
@SUM	This calculates the sum of the values in the specified range of cells.
@SUMSQ	This calculates the sum of the squares of the values in the specified range.
@SUMXMY2	This subtracts the values in range 2 from the values in range 1, squares the differences and then sums the results, being written in the form: @SUMXMY2(Range1;Range2) e.g. @SUMXMY2(B1:B7;C1:C7).
@VAR	This calculates the variance of a specified range.

In a formula the arithmetic operations are performed from left to right according to their order of precedence. The power operation ^ has the highest precedence and is performed first, then multiplication and division and finally addition or subtraction. Brackets are used to group operations. The expression in the leftmost brackets is evaluated first, using the precedence rule. Then the next brackets, until all such expressions have been evaluated. Then the resulting expression is evaluated from left to right. To illustrate the form formulas might take, consider the following examples. To take the mean value of the entries in cells B2 and C2, the formula might be entered as:

=(B2+C2)/2

The expression in the brackets is evaluated first and then the division carried out. To take the square of the value in cell B2 and divide it by twice the value in cell B6, the formula entered might be:

=B2^2/2B6

To evaluate h from the following equation (the height reached by a projectile with an initial velocity v at angle θ to the horizontal):

$$h = \frac{u^2 \sin\theta}{2g}$$

with the value for u in cell C2, that for θ in cell D2 and that for g in cell B1, the formula entered might be:

=C2^2*@SIN(D2*@PI/180)^2/(2*B1)

This uses special built-in functions which are specified by being preceded by @. In this instance the built-in function is π, denoted by PI. Thus, using the rules of precedence, the expression in the first brackets, i.e. (D2*@PI/180) is evaluated first, followed by @SIN(D2*@PI/180) and then (2*B1). Then C2^2 is evaluated, followed by @SIN(D2*@PI/180)^2, followed by C2^2*@SIN(D2*@PI/180)^2 and finally the complete expression of C2^2*@SIN(D2*@PI/180)^2/(2*B1) is evaluated.

7.1.3 Example of a completed spreadsheet

Consider a spreadsheet being used in an experiment for the determination of the acceleration due to gravity from a measurement of the periodic time of a simple pendulum. For a particular length L of pendulum 5 measurements were made of the times taken for 20 oscillations. The pendulum length was measured as 500 mm and the times obtained were 40.4, 40.6, 40.0, 40.6, 40.4 s. The acceleration due to gravity g is given by:

$$g = \frac{4\pi^2 L}{T^2}$$

where T is the periodic time, i.e. the time for 1 oscillation.

Figure 7.5 shows the spreadsheet with the formulas that could be used for the various cells. Row 1 has been used for a title, cells A2, B2 and C2 for the text length of a pendulum (m) and cell D2 for its value. Cell A3 has been used for the time symbol and its units and cells B3, C3, D3, E3 and F3 for the time values. Cells A4 and B4 are used for the text 'mean' and cell C4 for the formula for calculating the mean. Cells A5, B5, C5 and D5 are used for the text 'acceleration due to gravity' and cell D6 for the formula for calculating it. Note that the units must not be put in cells with values.

	A	B	C	D	E	F
1	Measurement of the acceleration due to gravity					
2	Length of pendulum (m)			0.500		
3	t(s)	40.4	40.6	40	40.6	40.4
4	Mean t (s)					
5	Acceleration due to gravity (m/s²)					
6						

Formula in C2: =(B3+C3+D3+E3+F3)/5
Formula in E5: =4*@PI^2*D2*20/C4^2

Figure 7.5 *Spreadsheet for acceleration due to gravity experiment*

With the formulas entered in the cells, when the data is entered the results automatically appear. If a measurement is changed, the results are automatically corrected to give the new value. If there is no entry in a cell it is treated as a zero. Figure 7.6 shows the spreadsheet of Figure 7.5 when the formulas are operative. We could have also included in the spreadsheet a formula for the calculation of the standard error.

	A	B	C	D	E	F
1	Measurement of the acceleration due to gravity					
2	Length of pendulum (m)			0.500		
3	t(s)	40.4	40.6	40	40.6	40.4
4	Mean t (s)		40.4			
5	Acceleration due to gravity (m/s²)				9.77	
6						

Figure 7.6 *Spreadsheet for acceleration due to gravity experiment*

7.1.4 Least squares method with a spreadsheet

As a further illustration of the use of a spreadsheet, consider the formulas required in order to carry out a least squares analysis of data. The data values are in columns B and C, extending from row 2 to row 11 as in Figure 7.7.

	A	B	C	D	E
1		x	y	xy	x^2
2					
3					
4					
5					
6					
7					
8					
9					
10					
11					
12	Sum				
13	m =				
14	c =				

Figure 7.7 *Spreadsheet for least squares*

The formula used in cell B12 to give the sum of the x values can be:

=@SUM(B2..B11)

Cell C12 is to have the sum of the y values and so might have the formula:

=@SUM(C2..C11)

Column D is to have the products of the corresponding values in columns B and C. Thus in cell D2 we might have the formula:

=B2*C2

In cell D3:

=B3*C3

and so on for the other product cells. Cell D12 is the sum of the xy values and so can have the formula:

=@SUM(D2..D11)

Column E has the squares of the corresponding x values. Thus the formula in cell E2 can be:

=B2^2

and in cell E3:

=B3^2

and so on. In cell E12 we have the formula for the sum of the x^2 values and so:

=@SUM(E2..E11)

Cell B13 contains the formula for the gradient m. The equation used to determine the gradient is (see section 6.1, equation [3]):

$$m = \frac{n\Sigma x_j^2 - \Sigma y_j \Sigma x_j}{n\Sigma x_j^2 - (\Sigma x_j)^2}$$

Thus the formula is:

=(10*E12−C12*B12)/(10*E12−B12^2)

Cell B14 contains the formula for the intercept c. The equation used to determine the intercept is (see section 6.1, equation [7]):

$$c = \frac{\Sigma x_j^2 \Sigma y_j - \Sigma x_j \Sigma x_j y_j}{n\Sigma x_j^2 - (\Sigma x_j)^2}$$

Thus the formula is:

=(E12*C12--B12*E12)/(10*E12−B12^2)

Thus cells B13 and B14 will contain the required answers.

7.2 Spreadsheets and charts

A range of types of charts can be produced using spreadsheets. These typically include: column, bar, X-Y, line and pie. In a *column chart*, the data points are displayed as rectangular columns with the height of a column representing the magnitude of the data. In a *bar chart*, the data points are displayed as rectangular bars with the length of a bar representing the magnitude of the data. An extension of the column and bar charts is the *stacked-column* and *stacked-bar* charts. In this type of chart, the column or bar corresponding to two or more values are stacked on top of each other. In an *X-Y graph*, the data points are plotted in the normal form of the Cartesian graph. A *line chart* is a form of *X-Y* graph with the abscissa values not equally spaced. The *pie chart*, as the name implies, is a circle divided into slices. Each slice of the pie corresponds to a data point, the angle subtended by the slice being proportional to the data value.

A typical sequence that is used to obtain charts from spreadsheet data is:

1. Select the range of cells to be used for the variables by holding down the mouse button and dragging the mouse from the first cell over the other cells to be included in the range. When the button is released, the edit line shows the range selected. Alternatively use the arrow keys to locate the first cell, then holding down the Shift key use the arrows to expand the range up or down or to the left or right.

2. Select Chart from the appropriate menu or click on the appropriate icon. Click and drag a box to where you want the chart to occur on the page. Lotus 1-2-3 will automatically, by default, create a bar chart using the selected data. If you want a different type of chart, select the Type from the chart menu and then select the required type from the drop-down menu. Various attributes of the chart can also be selected. In addition, the drop-down menu can be used to select headings, legends, i.e. explanatory notes, to be displayed, the fonts to be used for text, colours and tints to be used, and any criteria that are to be imposed for the ranges of axes. The axes are automatically set to encompass the range of the data; there may, however, be circumstances where you wish the range to be different.

3. View the chart.

The following examples show the types of chart that can be generated and the data used to generate them. Figure 7.8(a) shows the type of column chart and Figure 7.8(b) the type of bar chart that can be produced with the following data:

Length (mm)	12	13	14	15
Frequency (Hz)	3	6	4	1

The data shown below is of the temperature of a sample of water which is heated at a constant rate. Figure 7.9 shows the resulting *X-Y* graph.

Temperature °C	−20	0	0	0	10	20
Time (min)	0	4	8	12	16	20

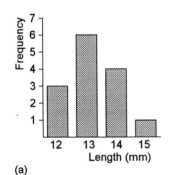

(a)

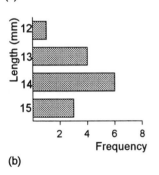

(b)

Figure 7.8 *(a) Column chart, (b) bar chart*

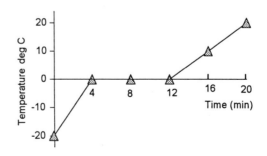

Figure 7.9 *X-Y plot*

The data below shows the breakdown of the number of hours spent by a student on a particular experiment and Figure 7.10 the resulting pie chart.

Planning and preparation (h)	2
Preliminary experiment (h)	1
Doing the experiment (h)	4
Analysis (h)	2
Report writing (h)	2

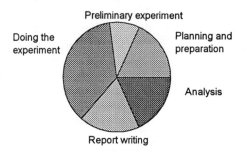

Figure 7.10 *Pie chart*

Problems 1 Devise formulas that could be used with spreadsheets to carry out the following operations:
(a) Adding the value in cell B3 to that in C3.
(b) Subtracting the value in cell D2 from that in B2.
(c) Multiplying the data in cell C5 by that in D5.
(d) Dividing the data in cell B4 by that in C4.
(e) Squaring the value of the data in cell C3.
(f) Multiplying the value in cell A4 by the square of the value in C2.
(g) Adding the values in cells B2, B3, B4, B5 and B6.
(h) Determining the mean value of the data in cells C2 and C3.
(i) Determining E using the equation $E = \frac{1}{2}mv^2$ when the value for m is in cell A5 and that for v in cell C3.
(j) Determine f using the equation:

$$\frac{1}{f} = \frac{1}{v} + \frac{1}{u}$$

where the value for v is in cell C2 and that for u in cell B2.
(k) Determine v using the equation $v = \sqrt{T/m}$, where the value for T is in cell B2 and that for m in cell C2.
(l) Determine the standard deviation of the squares of the deviations given for a sample of results in cells C2 to C11.

Appendix A: Experiments

The following are a few simple experiments that can be used to illustrate the points made in this book.

1. Determination of the acceleration due to gravity, using a simple pendulum

Aim
The aim of the experiment is to obtain a value for the acceleration due to gravity in the laboratory from measurements of the periodic time of a simple pendulum.

Theory
The periodic time T of a simple pendulum executing small angle oscillations is given by:

$$T = 2\pi \sqrt{\frac{L}{g}}$$

where L is the length of the pendulum and g the acceleration due to gravity. This equation can be rearranged to give:

$$L = \left(\frac{g}{4\pi^2}\right) T^2$$

Thus a graph of L against T^2 should be a straight line with a gradient of $(g/4\pi^2)$.

Apparatus
A simple pendulum consisting of a small, but dense, bob on the end of a length of thread; a stopwatch, metre rule, two small flat pieces of metal or wood, retort stand.

Procedure
Set up a pendulum of length about 40 cm. In order that the point of support can be accurately fixed, the end of the thread should be clamped between two flat pieces of metal or wood. Measure the length of the pendulum from the point of support of the thread to the centre of gravity of the bob. Determine the time taken for a suitable number of swings. The number of oscillations to be timed will depend on the accuracy required for the periodic time, the reading error of the stopwatch being the total time measured. Only a small angle of swing should be used, certainly less than about 5° from either side of the vertical. Repeat the time measurement a number of times and obtain a mean value.

Repeat the measurements for a number of different pendulum lengths and hence, by using a suitable graph or the least squares method, determine a value for the acceleration due to gravity.

Quote your result with an estimate of its accuracy and compare the value with that given in tables, discussing the significance of your result in the light of the quoted value and the possible sources of error in your experiment.

2. Determination of Young's modulus for the material of a metre rule

Aim

The aim of the experiment is to determine Young's modulus for the material of a metre rule from measurements of the depression of the free end when mounted as a cantilever and subject to loads applied at the free end.

Theory

For a beam mounted as a cantilever (Figure A.1), the depression y of the free end from the horizontal is given by:

$$y = \frac{4gL^3}{bd^3 E}(M + km)$$

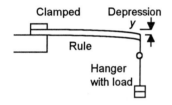

Figure A.1 *Cantilever*

where L is the length of the cantilever from the clamped point to the free end, g the acceleration due to gravity, b the breadth and d the thickness of the rectangular cross-section beam, M the mass of the applied load at the free end, m the mass of the beam, k a constant and E Young's modulus.

A graph of y against M should give a straight line graph with a gradient of $(4gL^3/bd^3E)$.

Apparatus

Metre rule for the beam, rule for measuring depression and retort stand to hold it in position, G-clamp to clamp metre rule to edge of a bench, weights (about 20 g are suitable) and hanger and a means of attaching the hanger to the end of the metre rule, e.g. a loop of wire, a large pin to use as a pointer and some means to attach it to the end of the rule, e.g. Sellotape.

Procedure

Using the G-clamp, fix the metre rule so that about 90 cm of its length overhangs the edge of a bench. Fasten the hanger to the free end of the metre rule and attach a large pin to the end to act as a horizontal pointer which can move against the vertical rule and so indicate the depression of the end. Note the initial position of the pointer and then add masses, say 20 g, to the hanger, noting the scale reading in each case. Measure the breadth and depth of the metre rule and the length that protrudes from the G-clamp.

Plot a suitable graph of the results, or use the least squares method, and so obtain a value for Young's modulus for the material of the metre rule. Assess the accuracy of your result. Look up the value of Young's modulus in tables and comment on the significance of your result in relation to the quoted value.

3. Determination of the resistivity of the material of a wire

Aim
The aim is to determine the resistivity of the material of a wire, the resistance of a length being measured by means of a metre bridge.

Theory
A length L of wire with a constant cross-sectional area A and an electrical resistance R has a resistivity ρ given by:

$$\rho = \frac{RA}{L}$$

The resistivity of metals depends on the temperature. Rearranging this equation gives:

$$R = \frac{\rho L}{A}$$

Thus a graph of resistance plotted against length of a wire should give a straight line graph with ρ/A as the gradient.

Apparatus
A length of uniform cross-section wire, e.g. 2 m of insulated Manganin of 30 S.W.G. or 2 m of constantan (Eureka) of 24 S.W.G., metre rule, micrometer screw gauge, metre bridge, 2 V battery for metre bridge, slider/jockey, centre-zero galvanometer, a resistance box, connecting wires, switch, thermometer.

Procedure
Assemble the circuit shown in Figure A.2. Connect a measured length of the wire in the left-hand gap in the bridge and the resistance box in the right-hand gap. Move the slider along the potentiometer wire until no current flows through the galvanometer. Record the distance x and use the following equation to determine the value of the resistance of the wire:

$$\frac{R_1}{R_2} = \frac{x}{100 - x}$$

Repeat the experiment for a number of different lengths of wire and by plotting a suitable graph, or using the least squares method, obtain a value for ρ/A. Record the temperature at which the measurement has been made.

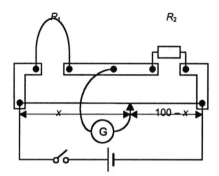

Figure A.2 *Metre bridge circuit*

Measure the diameter of the wire with a micrometer screw gauge, taking measurements at a number of positions along the wire and at each place in two directions at right angles. Before making the diameter measurements, the insulation must be carefully removed. Note any zero error in the micrometer screw gauge. Hence obtain a value for the resistivity, indicating the accuracy of your result and commenting on the sources of error in the measurements. Compare your result with that given in tables and comment on the significance of your result.

4. Determination of the viscosity of an oil

Aim
The aim is to determine the coefficient of viscosity of an oil from measurements of the terminal velocity with which a small sphere drops through the oil.

Theory
When a small sphere is dropped into a liquid, the velocity at first increases but, after a suitable length of fall, the viscous drag acting on the sphere becomes equal to the downward force on it and the velocity becomes constant, this constant value being termed the terminal velocity. The coefficient of viscosity η is related to the terminal velocity v by:

$$\eta = \frac{2}{9} r^2 \frac{(\rho - \sigma) g}{v}$$

where r is the radius of the sphere, ρ its density, σ the density of the oil and g the acceleration due to gravity. In deriving the above equation from Stoke's law, it has been assumed that the motion of the sphere is small so that no turbulence occurs and that the liquid is of infinite extent. In practice, the sphere cannot be dropped in a liquid of infinite extent and a correction has to be applied to allow for the effect of the walls of the containing vessel. A first approximation is the Ladenburg correction. This modifies the equation to:

$$\eta = \frac{2}{9}r^2 \frac{(\rho - \sigma)g}{(1 + 2.4d/D)v}$$

where d is the diameter of the sphere and D the diameter of the cylinder in which it is falling.

Apparatus
Glycerine or engine oil, steel balls of diameters in the range of about 3 mm to 6 mm, micrometer screw gauge, balance, hydrometer, stopwatch, metre rule, 1000 ml measuring cylinder or other large cylinder, thermometer, magnet to enable the steel balls to be removed from the oil. To investigate the correction for the diameter of the container, a number of tubes of different diameters which will fit within the measuring cylinder are required.

Procedure
Fill the measuring cylinder with the oil and allow it to stand long enough for all the air bubbles to escape. Measure the density of the oil by means of a hydrometer.

Select about 10 balls of the same diameter. Measure their diameters using a micrometer screw gauge. To obtain the density of the material used for the balls, weight all ten together and hence, from a value for the average mass, calculate the density, using the value of the volume computed from the diameter. Repeat this for the other diameter balls.

Release one of the spheres in the cylinder and measure the time it takes to fall between two graduation marks as far apart as possible. The terminal velocity in oil is reached after the ball has fallen through a distance of about six diameters, so the top mark should not be nearer to the upper surface than this. Repeat the timing for the other balls of the same diameter and so obtain a mean time for the fall. Measure the distance apart of the marks and hence deduce the mean velocity. Note the temperature of the oil. Repeat the experiment using another batch of spheres with a different diameter. Determine a value for the coefficient of viscosity by plotting a graph of v against r^2, or least squares analysis. Estimate the accuracy of your result.

To investigate the effect of the diameter of the container on the viscosity value obtained, repeat the experiment with tubes of different diameters. Then plot the value of the coefficient of viscosity obtained without correction against d/D and, from the shape of the resulting graph, consider the validity of the Ladenburg correction. Hence correct your results.

Discuss the accuracy of your measurement and the sources of possible error. Look up the value of the coefficient of viscosity in tables and compare it with your result, commenting on the significance of your result.

Appendix B: Further reading

The following are books that you might find useful in amplifying some of the points made in this book.

Anderson, H.L. editor (1989), *A Physicist's Desk Reference*, American Institute of Physics
   A reference book giving useful information, formulas, data, definitions and references relevant to physics.

Bevington, P.R. and Robinson, D.K. (1992), *Data Reduction and Error Analysis for the Physical Sciences*, 2nd Edition (McGraw-Hill)
   Useful for more detailed discussion of the statistics.

Callender, J.T. (1995), *Exploring Probability and Statistics with Spreadsheets*, Prentice Hall
   Very readable account of the use of spreadsheets in statistics.

Emden, J. van (1990), *A Handbook of Writing for Engineers*, Macmillan
   A booklet aiming to aid engineers in writing.

Jayaraman, S. (1991), *Computer Aided Problem Solving for Scientists and Engineers*, McGraw-Hill
   Use of computer programs, in particular spreadsheets.

Kaye, G.W.C. and Laby, T.H. (1986), *Tables of Physical and Chemical Constants*, 15th edition, Longman Scientific and Technical
   A reference book for physical, chemical and astronomical data.

Ross, S.M. (1987), *Introduction to Probability and Statistics for Engineers and Scientists*, John Wiley and Sons
   A more in-depth consideration of statistics.

Schenk, H. (1979), *Theories of Engineering Experimentation*, 3rd edition, Hemisphere Publishing Corporation
   Errors and statistics related to experimental work in engineering.

Squires, G.L. (1985), *Practical Physics*, 3rd edition, Cambridge University Press
   Similar coverage to this book and including basic laboratory instruments and methods in physics.

Also in relation to spreadsheets, books on the software available, e.g.

Dodge, M., Kinata, C. and Stinson, C. (1993), *Running Microsoft Excel for Windows*, Microsoft Press

Fein, M. and Sussman, A. (1994), *Teach yourself LOTUS 1-2-3*, MIS Press

Answers

Chapter 2
1. (a) 20×10^{-3} V, (b) 15×10^{-6} m^3, (c) 230×10^{-6} A, (d) 20×10^{-3} m^3, (e) 15×10^{-12} F, (f) 210×10^9 Pa, (g) 1×10^6 V
2. (a) 1.2 kV, (b) 0.2 MPa, (c) 200 dm^3, (d) 2.4 pF, (e) 3 mA, (f) 12 GHz
3. N s m^{-2}
4. J kg^{-1} K^{-1}
5. s^{-1} or Hz
6. (a) 13, (b) 0.21, (c) 0.014, (d) 19×10^2, (e) 0.0013
7. 26.2 g
8. 5.2×10^{-3} K^{-1}
9. a mV/K, b mV/K^2

Chapter 3
1. $\sigma = -0.025\theta + 85.2$
2. $n = 0.15V - 3$
3. $E = 0.21W + 10$
4. $E = 106$ kJ mol^{-1}, $A = 1.45 \times 10^{14}$ s^{-1}
5. $R = \dfrac{1200}{V} + 5$
6. $T = 2.0 L^{1/2}$
7. $Q = 2.6 h^{2.5}$
8. $v = 10.2\, e^{-0.1t}$
9. $V = -0.15t + 8$
10. $T = \dfrac{43.2}{s}$
11. 10 cm
12. $i = -0.1\, e^{-0.05t}$ mA
13. $\theta = 500\, e^{-0.2t}$
14. 127 mm

Chapter 4
1. ± 0.25°C
2. ± 0.05 units
3. (a) 50.2 s, 1.5 s, (b) 2.13 mm, 0.04 mm, (c) 50.2 cm^3, 2.2 cm^3
4. (a) 800 kN, (b) ± 0.07 kN
5. (a) 51.3 Ω, (b) $\pm 0.07\ \Omega$
6. (a) 39.0 kV, (b) ± 0.11 kV
7. $150 \pm 10\ \Omega$
8. $33.3 \pm 7.2\ \Omega$
9. $1.77 \pm 0.09 \times 10^6$ mm^2
10. 787 ± 24 kg/m^3
11. ± 0.6 g
12. 60 ± 2 km
13. $2.0 \pm 0.4 \times 10^5$ mm^3
14. 0.400 ± 0.046 kN/mm^2

15 $\Delta s = \frac{1}{2} c \sin A . \Delta b + \frac{1}{2} b \sin A . \Delta c + \frac{1}{2} bc \cos A . \Delta$
16 $\Delta N = e^{-\lambda t} \Delta N_0 + N_0 \lambda \, e^{-\lambda t} \Delta t$
17 $\Delta \eta = \frac{2}{9} g (\rho_s - \rho_l) \left(\frac{2r}{v} \Delta r + \frac{r^2}{v^2} \Delta v \right)$

Chapter 5
1 40, 31.6
2 (a) 0.02, 0.06, 0.22, 0.32, 0.28, 0.08, 0.02, (b) 69.3, 2.3
3 0.5
4 20
5 $22.35 \leq \bar{x} \leq 27.47$
6 $19.33 \leq \bar{x} \leq 20.27$ s
7 $18.45 \leq \bar{x} \leq 20.51$ Ω
8 $12.1 \leq \bar{x} \leq 16.1$ µF
9 Yes
10 No, set too high
11 No
12 Yes, reading high
13 Could be the same
14 Yes
15 Unlikely
16 Highly significant
17 No
18 Yes

Chapter 6
1 $y = 0.66x + 29.15$
2 $e = 6.0W + 5.6$
3 $m = 0.500\theta + 44.86$
4 $d = 9.884W + 0.062$
5 $y = 0.857x + 0$
6 $y = 1.538x + 0.640$
7 $E = 0.307W + 1.567$
8 $y = 3.071 \, e^{0.5056x}$
9 $y = 0.40x^{1.70}$
10 $y = 1.044 \, e^{0.682x}$
11 20.1286 ± 0.0015 Ω
12 9.8107 ± 0.0005 m/s^2

Chapter 7 1 (a) =B3+B4, (b) =B2-D2, (c) =C5*D5, (d) =B4/C4, (e) =C3^2, (f) =A4*C2^2, (g) =@SUM(B2:B6), (h) =(C2+C3)/2, (i) =(A5*C3^2)/2, (j) =B2*C2/(B2+C2), (k) =@SQRT(B2/C2), (l) =@SUM(C2:C11)/9

Index

Acceleration due to gravity:
    experiment, 127
    records, 5
    report, 4
    spreadsheet, 121
Accuracy, defined, 46
Area under graph, 40
Average, *see Mean*

Bar chart, 124, 125

Charts, 124
Column chart, 124, 125
Confidence interval, 86

Degree of freedom, 89
Deviation:
    from mean, 52
    standard, 52, 75
Distribution:
    frequency, 70
    Gaussian, 84
    limiting frequency, 71
    normal, 84
    normalised, 71
    Student's t, 88
    t, 88

Error:
    bars on graphs, 26
    combining, 56
    defined, 46
    fractional, 47
    human, 48
    insertion, 48, 49
    loading, 49
    non-linearity, 48
    parallax, 48
    percentage, 47
    random, 51
    reading, 48, 49
    sources, 47
    straight line graphs, 32, 103
    systematic, 51
Experiments:
    errors, 46
    examples, 127
    record keeping, 3
    report writing, 4
    stages, 1
Exponent, 15
Extrapolation, 30

Frequency:
    defined, 70
    relative, 70

Gaussian distribution, 84
Graphs:
    area, 40
    drawing, 26
    error bars, 26
    extrapolation, 30
    interpolation, 30
    least squares line, 97
    line of best fit, 27
    linear scale, 25
    logarithmic scale, 25, 37
    plotting, 24
    straight line, 30

Histogram, 70

Interpolation, 30

Laboratory:
    diary, 3
    journal, 3
    notebook, 3
Least squares:
    error, 103
    method, 97
    spreadsheet, 122
    weighted, 110
Light, speed of, example, 73

Line chart, 124
Linearising equations, 33
Loading error, 49

Mean:
   defined, 51, 72
   deviation from, 52
   difference between two, 91
   error of, 54
   of distribution, 73
   of histogram, 73
   standard error, 55, 60, 66, 81
   weighted, 107
Mid-ordinate rule, 40

Normal distribution, 84
Normalised distribution, 71

Parallax, 48
Partial differentiation, 64
Pie chart, 124, 126
Powers of ten, 15
Precision, 75
Probability, 71

Random errors, 51
Record keeping, 3
Report writing, 4
Resistivity experiment, 129
Rounding, 20

Scientific notation, 15
SI units, 11
Significance, 86
Significant figures, 19
Spreadsheet:
   cell addresses, 116
   charts, 124
   control panel, 116
   data in cell, 117
   described, 116
   edit line, 116
   example, 121
   formula in cell, 118
   formula specification, 118
   least squares, 122
   text in cell, 117
Standard:
   deviation, 52, 75
   deviation, pooled, 92
   error, 55, 60, 66, 81
   notation, 15
Straight line graphs, 30
Student's t-distribution, 88
Style, writing, 8
Systematic errors, 51

t-distribution, 88
Tables, data in, 18
True value, 51, 73

Units:
   equations, 13
   non-SI, 18
   prefixes, 16
   SI, 11

Variance, 53, 76
Viscosity experiment, 130
Voltmeter, loading error, 49

Weighting, 107

X-Y chart, 124, 125

Young's modulus experiment, 128